T0226672

THE CHOLESTEROL WARS

To the hundreds of scientists, clinicians and associated research personnel who over many years amassed the evidence that hypercholesterolemia is a key causative factor in atherosclerosis; and to the thousands of volunteers who agreed to participate in the critically important clinical trials that closed the case, inaugurating a new era in preventive cardiology.

THE CHOLESTEROL WARS

The Skeptics vs. the Preponderance of Evidence

Daniel Steinberg, M.D., Ph.D.,
University of California, San Diego

AMSTERDAM • BOSTON • HEIDELBERG • LONDON • NEW YORK • OXFORD
PARIS • SAN DIEGO • SAN FRANCISCO • SINGAPORE • SYDNEY • TOKYO
Academic Press is an imprint of Elsevier

Academic Press is an imprint of Elsevier
525 B Street, Suite 1900, San Diego, California 92101-4495, USA
360 Park Avenue South, New York, NY 10010-1710, USA
84 Theobald's Road, London WC1X 8RR, UK
30 Corporate Drive, Suite 400, Burlington, MA 01803, USA

First edition 2007

Copyright © 2007 Elsevier Inc. All rights reserved

No part of this publication may be reproduced, stored in a retrieval system
or transmitted in any form or by any means electronic, mechanical, photocopying,
recording or otherwise without the prior written permission of the publisher

Permissions may be sought directly from Elsevier's Science & Technology
Rights Department in Oxford, UK: phone (+44) (0) 1865 843830;
fax (+44) (0) 1865 853333; email: permissions@elsevier.com. Alternatively
you can submit your request online by visiting the Elsevier web site at
http://elsevier.com/locate/permissions, and selecting
Obtaining permission to use Elsevier material

Notice
No responsibility is assumed by the publisher for any injury and/or damage to persons
or property as a matter of products liability, negligence or otherwise, or from any use
or operation of any methods, products, instructions or ideas contained in the material
herein. Because of rapid advances in the medical sciences, in particular, independent
verification of diagnoses and drug dosages should be made

British Library Cataloguing in Publication Data
A catalogue record for this book is available from the British Library

Library of Congress Cataloging in Publication Data
A catalog record for this book is available from the Library of Congress

ISBN: 978-0-12-373979-7

For information on all Academic Press publications
visit our web site at books.elsevier.com

Printed and bound in Great Britain by
CPI Antony Rowe, Chippenham and Eastbourne

Transferred to Digital Printing, 2011

Working together to grow
libraries in developing countries

www.elsevier.com | www.bookaid.org | www.sabre.org

ELSEVIER   BOOK AID
International   Sabre Foundation

CONTENTS

Contents

Contents

FOREWORD

The bravest are surely those who have the clearest vision of what is before them.
Thucydides, ~404 BC

The tradition of chronicling wars while they are being fought was initiated in the 5th century BC with the description of the Peloponnesian Wars by Thucydides, a general in the Athenian army. That noble tradition is extended in elegant fashion in this book by Daniel Steinberg, a general in the Cholesterol Wars. The Cholesterol Wars began nearly 100 years ago in Russia when a young pathologist, Nikolai Anitschkow, fed cholesterol to rabbits and produced atherosclerosis of the arteries. This experiment and subsequent epidemiologic studies triggered a passionate debate as to whether cholesterol is the root cause of human atherosclerosis, the disease process that underlies heart attacks and strokes. The stakes were high. Throughout the 20th century, while the cholesterol battles were raging, more people were dying of atherosclerosis than were killed in military combat.

Although the Cholesterol Wars are not over, major battles have been won by the anti-cholesterol forces, i.e. those who condemn cholesterol as the culprit. Like modern armies, the anti-cholesterol forces have been aided by powerful new weapons: (1) profound insights into the mechanisms by which lipoproteins such as low density lipoprotein (LDL) and high density lipoprotein (HDL) transport cholesterol in the blood; (2) unmasking of the regulatory mechanisms that control these processes; (3) molecular delineation of genetic factors that elevate blood LDL-cholesterol and accelerate atherosclerosis; and (4) development of relatively safe and effective drugs that lower LDL-cholesterol and reduce heart attacks. Steinberg replays all of these battles in succinct and compelling fashion. He names the protagonists and the antagonists. He supplies delicious quotations from skeptics of the "cholesterol hypothesis". He quotes an eminent British epidemiologist who described English skepticism as follows: "Cholesterol was something that Americans had; it definitely wasn't British. Anyone who talked about cholesterol was obviously suffering from American-style hypochondria." (see p. 8). The speaker was characterizing attitudes as recent as the 1990s.

While these battles were being fought, public health authorities in the U.S., U.K., and elsewhere hesitated to make the general recommendation of aggressive cholesterol lowering to society as a whole. This inexplicable delay resulted in widespread public confusion and cost many thousands of lives. For the anticholesterol forces, it is still too early to declare "mission accomplished". Although mortality from heart attacks and strokes has been reduced, these catastrophes are still major killers.

Where is the next battleground? We agree with Steinberg that the next battle will be fought over the issue of when to start LDL-lowering therapy. All of the epidemiologic data suggests that "the earlier the better" (see Chapter 9). This conclusion has been reinforced recently with the discovery of a common genetic trait in African-Americans that lowers blood LDL-cholesterol levels by only 28 percent throughout life, yet leads to an 88 percent reduction in coronary heart disease in the sixth and seventh decades despite contributory "risk factors" that include hypertension and diabetes. This 88 percent reduction in coronary events is much greater than the 30–40 percent reduction that is attained when LDL-cholesterol levels are lowered by diet or drugs that are started many years after the atherosclerotic process has developed.

Autopsy studies, such as those on young men who died in the Korean War, have documented that the earliest hallmarks of atherosclerosis begin before 20 years of age. While few would suggest aggressive LDL lowering for most teenagers, there must be a time between age 20 and 40 when it is appropriate to begin careful attention to plasma LDL-cholesterol in everyone. The standard intervention is a low-cholesterol, low-fat diet that is relatively rich in polyunsaturated fatty acids. When consumed faithfully over a lifetime, as it was in China and Japan before recent times, such diets lower plasma LDL-cholesterol sufficiently to reduce the incidence of heart attacks by more than 90 percent. Yet, as Steinberg suggests, it would take decades to change the dietary habits of Americans sufficiently to make such an impact. Moreover, the economic and social consequences would be great. Thus, it is likely that we will see more widespread use of LDL-lowering drugs such as statins and cholesterol absorption inhibitors. All evidence predicts that these drugs would lower markedly the heart attack burden if they were started early enough. Yet, as Steinberg documents, the remaining skeptics still challenge the wisdom of such widespread use.

When a scholarly book like this one is published, the question always arises as to who should read it. In our opinion, *The Cholesterol Wars: The Cholesterol Skeptics vs. the Preponderance of Evidence* should certainly be read by anyone who is in a position to influence public policy toward health issues. It should also be read by all physicians who care for patients at risk for heart disease or stroke. Finally, it should be read by members of the general public who are puzzled by the conflicting claims about cholesterol that continue to be made. Armed with the facts outlined in this book, any individual will be fully equipped to stake out an informed position in the next Cholesterol War.

Michael S. Brown and Joseph L. Goldstein
University of Texas Southwestern Medical Center
Dallas, Texas, USA

PREFACE

The thesis of this book is that the importance of hypercholesterolemia in human atherosclerosis should have been and could have been appreciated decades earlier than it was. The opportunities that were missed and the findings that went unappreciated because of preconceived mindsets are reviewed and analyzed. The history of the controversy is intrinsically of interest. In addition, there may be lessons to be learned from that history that could provide guidance in dealing with controversies yet to come.

This book is largely confined to the key events that ultimately established dyslipidemia (elevated blood levels of low density lipoprotein or/and low blood levels of high density lipoprotein) as causative in atherosclerosis. Of course, many factors in addition to dyslipidemia contribute to the atherogenic process and help determine when and how aggressively it will express itself clinically. Cigarette smoking, hypertension, obesity, diabetes, and family history are among the critical ones. No attempt is made here to deal with these in any detail except as they bear on the evolution of the lipid hypothesis. That should not in any way be taken as a comment on their importance. They are also critically important. Nor does this book attempt to present a broad review of atherogenesis research or of lipid and lipoprotein research, again, except as they fed into the development and validation of the lipid hypothesis. The book is primarily an inquiry into how, after much controversy, cholesterol and lipoproteins were indicted, tried, and ultimately found guilty as major contributors to the development of the atherosclerotic lesion and its clinical consequences.

Having started my medical training in 1941 and having done research relating to atherosclerosis for more than 40 years, I have lived through the "cholesterol wars" myself – and have a few scars to show for it. In addition to carrying on my own clinical and laboratory research programs, I have always tried to do my part in national cooperative programs and policy-making committees. Specifically, the reader should know that I have been actively involved in both the research and the public policy developments on the cholesterol front. I was chairman of the committee that designed the landmark clinical trial that, in 1984, demonstrated that lowering blood cholesterol levels with cholestyramine significantly decreased coronary heart disease risk (the Lipid Research Clinics Coronary Primary Prevention Trial) (1–3). I was also chairman of the 1984 Consensus Conference on Lowering Blood Cholesterol to Prevent Heart Attacks, sponsored by the National Institutes of Health (NIH) (4) and a member of the first Expert Panel asked by the National Cholesterol Education Program to issue guidelines for managing high blood cholesterol (5). On the positive side, this involvement means that I know first hand much of the scientific background and many of the players involved in this history. On the negative side, my close involvement with the events as they unfolded could have led to biases that help shape how I tell the history. I have tried my best to be objective but that is sometimes hard for warriors and I may not always have been successful. For errors, omissions, or imbalances that have crept in I apologize in advance.

Portions of this book first appeared in five excerpts published as invited reviews in the *Journal of Lipid Research*, a publication of the American Society for Biochemistry and Molecular Biology (6–10). The author is deeply indebted to the editors, Edward A. Dennis and Joseph L. Witztum, for inviting those reviews and to the Society for generously giving copyright permission for the inclusion of some of that material in this book.

The intended primary audience for the book is the biomedical community. However, there may be material here of interest to people in other fields. For that reason a glossary is appended that may be helpful to them.

This book might never have been written if Irvine H. Page had not suggested it to me almost 20 years ago. He wrote me a note shortly before his death in 1991, saying in essence that it was my duty "to write the history – so that people will remember the uphill battle it was to gain acceptance of the lipid hypothesis." Page was one of the first cardiologists to recognize the plausibility of the hypothesis and to champion efforts to prove it at the clinical level (11;12). He and his long-time collaborator, Lena A. Lewis, published, in 1969, an article with the ingenuous title: "A long-time study of the blood lipids of two students of atherosclerosis" (13). In it he reveals that he himself had a total blood cholesterol over 300 mg/dl when untreated. With a diet low in saturated fat and high in polyunsaturated fat he managed to keep his cholesterol level between 230 and 270 mg/dl, which is not bad, but would hardly be considered satisfactory treatment by today's standards. He had a strong family history of cardiovascular disease on his father's side and, in June 1967, he had a myocardial infarction. He died 24 years later, 5 months after his 90th birthday. Whether that was because of his diet or in spite of it, we will never know. In one of his many charming discourses on the subject of atherosclerosis he acknowledged that the lipid hypothesis was at that time (late 1960s) not yet solidly proved. He went on, however, to say that he personally was following a heart-healthy diet anyway – because he "didn't want to be the smartest man in the cemetery." Irv, thanks for your mandate to write the history.

REFERENCES

1. Steinberg, D. 1975. Planning the type II coronary primary prevention trial of the lipid research clinics (U.S.A.). *Adv Exp Med Biol* 63:417–426.
2. 1984. The Lipid Research Clinics Coronary Primary Prevention Trial results. I. Reduction in incidence of coronary heart disease. *JAMA* 251:351–364.
3. 1984. The Lipid Research Clinics Coronary Primary Prevention Trial results. II. The relationship of reduction in incidence of coronary heart disease to cholesterol lowering. *JAMA* 251:365–374.
4. 1985. Consensus conference. Lowering blood cholesterol to prevent heart disease. *JAMA* 253:2080–2086.
5. 1988. Report of the National Cholesterol Education Program Expert Panel on Detection, Evaluation, and Treatment of High Blood Cholesterol in Adults. The Expert Panel. *Arch Intern Med* 148:36–69.
6. Steinberg, D. 2004. An interpretive history of the cholesterol controversy, Part I. *J Lipid Res* 45:1583–1593.
7. Steinberg, D. 2005. An interpretive history of the cholesterol controversy, Part II: The early evidence linking blood cholesterol to coronary disease in humans. *J Lipid Res* 46:179–190.
8. Steinberg, D. 2005. An interpretive history of the cholesterol controversy, Part III: Mechanistically defining the role of hyperlipidemia in the pathogenesis. *J Lipid Res* 46:2037–2051.
9. Steinberg, D. 2006. An interpretive history of the cholesterol controversy, Part IV: The 1984 Coronary Primary Prevention Trial ends it – almost. *J Lipid Res* 47:1–14.
10. Steinberg, D. 2006. An interpretive history of the cholesterol controversy, Part V: The discovery of the statins and the end of the controversy. *J Lipid Res* 47:1339–1351.
11. Page, I.H., Stare, F.J., Corcoran, A.C., Pollack, H., and Wilkinson, C.F., Jr. 1957. Atherosclerosis and the fat content of the diet. *Circulation* 16:163–178.
12. Page, I.H. 1968. Atherosclerosis. A personal overview. *Circulation* 38:1164–1172.
13. Page, I.H. and Lewis, L.A. 1969. A long-time study of the blood lipids of two students of atherosclerosis. *Circulation* 40:915–918.

ACKNOWLEDGMENTS

I have enjoyed unstintingly generous cooperation and advice from every single colleague I turned to during the preparation of this history ... and they were many. My thanks to each and every one of them.

Special thanks go to some of the major players in the history who made themselves available for interviews either in person, by telephone or in correspondence. They included:

Albert W. Alberts
Michael S. Brown
James Cleeman
Barry Collins
Akira Endo
Donald S. Fredrickson
John W. Gofman
Joseph L. Goldstein
Anatoli Klimov
Claude Lenfant
David G. Orloff
Sir Richard Peto
Basil Rifkind
Solomon Sobel
P. Roy Vagelos
Joseph L. Witztum.

All of those on the list above kindly critiqued at least the sections of the book in their areas of expertise. A number of other people in the field also took the time to read sections of the book and offered useful suggestions. They included:

Henry Blackburn
William E. Connor
Godfrey S. Getz
Scott M. Grundy
Richard J. Havel
John Kane
Olga Stein
Yechezkiel Stein.

I have almost without exception gone to primary sources and need to thank Anatoly Klimov and Yury Miller for translations from the Russian; Valeska Terpstra for translations from the Dutch; and Wulf Palinski and Oswald Quehenberger for translations from the German.

J. Elliott in the Office of Medical Research applications at NIH kindly provided me with the taped proceedings of the 1984 Consensus Conference on Lowering Blood Cholesterol to Prevent Heart Disease.

Jean D. Wilson and Daniel W. Foster helped greatly with background material on their colleagues at the University of Texas Southwestern Medical Center at Dallas, Joseph L. Goldstein and Michael S. Brown.

I owe a lot to Jasna Markovac, Senior Vice-President at Elsevier, for seeing enough promise in my proposal to offer me a contract, and to Tari Broderick, Senior Publishing Editor at Elsevier, who helped me thread my way through the intricacies of the publishing world. Melissa Read and Deena Burgess handled production deftly. Let me also recognize the invaluable help of the research staff at the National Library of Medicine in Bethesda and at the Biomedical Library of the University of California San Diego. These so essential people too often go unsung.

Many thanks to Arlyne Lazerson, freelance editor and former book editor at *Psychology Today*, for her tender loving care in combing my manuscript for snarls. And to Oswald Quehenberger, Richard Elam, and Maureen Hargrave for taking me by the hand and leading me through the minefields of Word, Powerpoint, and Adobe Photoshop.

Without the support and encouragement of my dear wife, Mary Ellen Stratthaus, I might not have made it. Thanks Emmy.

My eternal gratitude goes to my good friend and long-time collaborator, Joseph L. Witztum, who provided me with space and support for this post-Emeritus undertaking. Some of the themes in the book may have been drawn from his songbook. We sang them together for many years.

Finally thanks to so many of my colleagues at UCSD with whom I have enjoyed discussing (and sometimes fuming about) the Great Cholesterol Controversy over the past 50 years.

CHAPTER 1

Overview

This book is an analysis of the controversy that swirled about the "lipid hypothesis" during the latter half of the 20th century. The "lipid hypothesis" postulates that hypercholesterolemia is a major causative factor in atherosclerosis and coronary heart disease. It does not propose that hypercholesterolemia is the *only* causative factor; certainly cigarette smoking, hypertension, diabetes, obesity, local arterial abnormalities, and probably a number of other factors are also highly relevant. What the hypothesis does propose is that hypercholesterolemia is what might be called a *determining factor*, i.e. it is sufficiently dominant that correcting it will significantly reduce the burden of disease and its clinical consequences even if hypercholesterolemia is the sole variable manipulated.

Note also that the hypothesis relates to *blood* lipids, not *dietary* lipids, as the putative directly causative factor. Although diet, especially dietary lipid, is an important determinant of blood lipid levels, many other factors play important roles. Moreover, there is a great deal of variability in the response of individuals to dietary manipulations. Thus, it is essential to distinguish between the indirect "diet–heart" connection and the direct "blood lipid–heart" connection. Failure to make this distinction has been a frequent source of confusion.

ON THE NATURE OF MEDICAL CONTROVERSY

The history of science is studded with controversies and perhaps this is especially true in medical science. For one thing, acceptance of a new hypothesis in medicine may entail introduction of new treatment modalities. If so, the decision to act on the assumption that the hypothesis is correct may have life and death consequences. For that reason some measure of skepticism is appropriate. New hypotheses should be critically tested and new treatments should be carefully evaluated for efficacy and safety. Some degree of conservatism is a

The Cholesterol Wars by D. Steinberg.
Copyright © 2007 Elsevier Inc. All rights of reproduction in any form reserved.

virtue when it prevents the hasty adoption of inadequately tested new treatments (e.g. the use of thalidomide in women of child-bearing age). On the other hand, an overly cautious approach and an exaggerated skepticism can delay the introduction of therapies that might save many lives. Atherosclerosis and its complications are responsible for about 500,000 deaths annually in the United States alone; it slightly outranks even cancer as a cause of death. Delaying the introduction of an effective intervention for even a few years may have had enormous consequences. This book reviews the scientific advances that built the case for the lipid hypothesis. In parallel, it examines the possible reasons for the exaggerated level of skepticism in the biomedical community. Perhaps there are lessons to be learned that will help us avoid making the same kinds of mistakes in the future.

The correctness of the lipid hypothesis is now firmly established but getting to this point has not been easy. Those who rejected the hypothesis were highly vocal in their opposition. Even when they represented a small minority of scientists in the field, they were able, especially in the United Kingdom, to hold back progress toward implementing screening and public health programs to lower blood cholesterol. Were the naysayers unqualified? No, most were respected and experienced in their fields. Were they looking at a different data set? No, the same data in the literature were available to all. Were they concerned about the danger of interventions to lower cholesterol levels? To some extent perhaps but there was no good evidence for such a danger.

As we recount the advances in knowledge that ultimately proved the correctness of the hypothesis, we will examine the criticisms offered by the opponents. There were several, but perhaps the major difference between the "convinced" and the "unconvinced" was that the latter were unwilling to look at *the totality of the evidence*, i.e. to evaluate the related but relevant separate contributions coming from different sources – from experimental animal studies, from epidemiologic studies, from genetic analyses, and from clinical observations. Each of these fields, approaching the problem from different directions, contributed solid evidence strongly implicating hypercholesterolemia as a major causal factor in atherosclerosis and its clinical expression. And they did so well before definitive clinical trials made the causal connection clear and unarguable. However, to the naysayers, any perceived weakness in any one line of evidence or any one clinical trial was considered sufficient to justify dismissing the hypothesis out of hand – and they were quick to point to anything that did not fit. It is true that in assessing the validity of a deductive hypothesis (*if a then b*), even a single discordant observation (*a and not b*) refutes the hypothesis. Such an uncompromising standard for refutation may be warranted, say, in the case of basic theories in physics, but that is not the case in medical science, which is only now inching its way toward membership in the hard sciences. In medicine we still deal with *statistical hypotheses* (*if a then b with some probability c*). Occasional discordant observations (*a and not b*) certainly weaken the proposed causal connection but do so only with some degree of probability, not absolutely.

How are we to "look at all of the evidence"? When we deal with data generated from identical or very similar protocols, as in closely related clinical intervention trials, we can combine the results using meta-analysis. Essentially that lets us take advantage of the much larger numbers represented when data from all the available studies that fit the predefined category are pooled. It's like having a much larger "n" when deciding what is statistically significant. Unfortunately, we do not yet have analogous methods for pooling evidence of qualitatively different kinds. For example, we know that atherosclerosis can be produced in almost any experimental animal if its blood cholesterol level can be raised high enough, strongly suggesting that the same may be true in another animal species – *Homo sapiens*. Of course, it is always a possibility (however slim) that humans might be an exception. How do we assign a number to that probability? Coronary heart disease risk in different countries parallels the level of blood cholesterol in those countries, but how much weight should we attach to that observation? Patients with low density lipoprotein (LDL) receptor deficiency have extremely high cholesterol values and have strikingly premature coronary heart disease, but how do we assign a number that lets us balance that against a negative clinical trial in a general population using a cholesterol-lowering regimen? Clearly all these lines of evidence are relevant to the issue of causality yet none of them individually may be able to establish it conclusively.

Suppose we run a clinical intervention trial to see if lowering blood cholesterol reduces coronary heart disease risk. Suppose the trial shows a 10 percent reduction in cholesterol level, a 20 percent reduction in risk, and a p-value of 0.08. By conventional criteria the results of the trial are not "statistically significant," i.e. p is greater than 0.05. Let us suppose further that there are no funds to support additional clinical trials. As a member of an NIH committee you are asked to advise whether the treatment should be offered to individuals with profiles like those of the subjects in the trial. Your decision might be affected if there were additional lines of evidence supporting causality even though the trial itself was not "statistically significant." You might very well decide that *the totality of the evidence* justified action despite the borderline significance of the trial data. This is the kind of situation in which the "pooling" or "merging" of different lines of evidence may be essential in reaching the right conclusion. This is hardly a new concept. Thomas Bayes, an 18th century British statistician, proposed that in evaluating any given hypothesis it is essential to consider not only the probability emerging from the immediate test (e.g. a randomized clinical trial of a new treatment modality) but also all of the previous findings bearing on the hypothesis (1;2). Basically he was saying that we should try to take into account all antecedent findings along with the probability derived from the immediate trial under evaluation.

AN OVERVIEW OF THE CHOLESTEROL WARS

To some younger colleagues it may come as a surprise that hypercholesterolemia has not always been accepted as a major factor in atherosclerosis and coronary heart disease. Yet, in 1946, Peters and Van Slyke, in their classic textbook, *Quantitative Clinical Chemistry*, concluded that "although there can be no doubt that deposits of lipids, especially cholesterol, are consistent and characteristic features [of atherosclerotic lesions] there is no indication that hypercholesterolemia plays more than a contributory role in their production" (3). In 1946, most physicians still considered atherosclerosis to be an inevitable accompaniment of aging about which nothing could be done. In fact, nothing much *was* being done, despite the accumulated evidence that strongly supported the lipid hypothesis.

ATHEROSCLEROSIS IN EXPERIMENTAL ANIMALS (CHAPTER 2)

The fact that rabbits could be made hypercholesterolemic just by feeding them pure cholesterol and that their arteries then showed lesions closely resembling those of human atherosclerosis was beautifully demonstrated in 1913 by Anitschkow and Chalatov (4). Over the following decades it was shown that atherosclerosis can be produced in virtually any experimental animal provided a way can be found to elevate the blood cholesterol sufficiently.

HYPERCHOLESTEROLEMIA AS A CAUSATIVE FACTOR IN THE HUMAN DISEASE (CHAPTER 3)

Can the findings from experimental atherosclerosis in animals be extrapolated to humans? Yes. In 1939, an astute clinician in Oslo, Carl Müller, pulled together many case reports on the rare syndrome of hereditary xanthomatosis and concluded that the marked hypercholesterolemia in these cases was the cause of both the skin lesions and the lesions in the coronary arteries (5).

Can the findings in these rare and extreme examples of hypercholesterolemia be extrapolated to the general population? Yes. The classic studies of Keys *et al.* (6) and the NIH-sponsored Framingham study (7) showed that coronary heart disease risk was proportional to blood cholesterol levels in the population at large, i.e. *not* limited to those with rare genetic defects.

Can anything be done to lower an elevated blood cholesterol? Yes. By the early 1950s it had been established that blood cholesterol levels in humans could be experimentally lowered by substituting dietary polyunsaturated fats for saturated fats as a single variable in metabolic ward studies (8;9).

Can lowering blood cholesterol level reduce the progression and clinical expression of atherosclerosis? Yes. In 1966, Paul Leren published his classic 5-year study of 412 patients who had had a prior myocardial infarction. He showed that substitution of a polyunsaturated fat for the saturated fat-rich, butter-cream-venison diet favored by the Norwegians reduced their blood cholesterol by about 17 percent and kept it down. The number of secondary coronary events in the treated group was reduced by about one-third and the result was significant at the $p < 0.03$ level (10). Over the ensuing decades the case against hypercholesterolemia became ever stronger, including the publication of two additional large-scale dietary intervention trials showing significant reduction in coronary events or stroke (11;12).

THE FIRST STEPS TOWARD IMPLEMENTATION OF A CHOLESTEROL-LOWERING STRATEGY

On the basis of these emerging findings, the American Heart Association as early as 1961 had already accepted the causal relationship and recommended that people at high risk be advised to modify their diets to avert heart attacks (13). In 1964, they extended their dietary recommendations to the general public and, in 1965, the Food and Nutrition Council of the American Medical Association made similar recommendations. In 1969, the Chairman of the Council on Arteriosclerosis of the American Heart Association said in his Presidential Address: "It is now good medical practice to treat – and I use the word advisedly – people who have definite hyperlipoproteinemia" (14). However, few practitioners were paying any attention to hypercholesterolemia; dietary advice was minimal and drug treatment was still in its infancy. The situation in 1968 was succinctly described by I.D. Frantz, Jr. and Richard B. Moore (15):

> Few controversies have divided the medical community so sharply for such a long time as has the sterol hypothesis. The separation between the two points of view has become so extreme that, on the one hand, there are respected scientists who believe that the evidence is already so convincing that further clinical testing is unnecessary, financially wasteful and actually unethical; and, aligned against them, are equally respected scientists who believe that the total weight of evidence accumulated over the many years is too slight to justify further work along these lines.

THE BASIC SCIENCE FOUNDATION (CHAPTER 4)

Preceding and paralleling the early clinical trials there were important advances being made in the basic sciences relevant to understanding the pathobiology of atherosclerosis and moving toward its prevention. These included importantly the elucidation of the pathway for biosynthesis of cholesterol (16) and the recognition of the heterogeneity of the lipoproteins and their complex metabolism (17). By the 1970s, the schema for lipoprotein metabolism was fairly well worked out and many of the genetically determined dyslipidemias had been well characterized (18–22).

PATHOGENESIS (CHAPTER 5)

Reluctance to accept the lipid hypothesis may have been based in part on the uncertainties still surrounding its pathogenesis. The groundbreaking work of Ross and Glomset brought a new level of sophistication to the field, probing the cellular and molecular events in the developing lesion (23;24). Goldstein and Brown's discovery of the LDL receptor gene as the specific mutation in familial hypercholesterolemia was a major milestone (25;26). Since familial hypercholesterolemia was known to be a monogenic disorder it was now possible to say that the premature atherosclerosis and coronary heart disease in these patients had to be ascribed to the high LDL levels induced by mutation of the LDL receptor gene.

The cellular basis for cholesterol accumulation in macrophages remained unclear until Brown and Goldstein discovered the macrophage scavenger receptors, receptors that recognize modified forms of LDL but not native LDL (27;28). Several biological and chemical modifications were shown to generate modified forms of LDL that were specifically recognized by one or more scavenger receptors (29–33). In parallel with these studies there was an explosion of interest in the role of the immune system in atherogenesis and in extensions of Ross's concept of atherosclerosis as an inflammatory disease. Interest in that aspect of atherogenesis continues to this day (34–36).

Although the dietary intervention trials had implicated hypercholesterolemia as causal, the evidence from the early trials themselves was not strong and many physicians remained unpersuaded. What was needed was some way to lower cholesterol levels more effectively. A few drugs that worked were already available but each had important drawbacks. Nevertheless, several clinical drug trials were undertaken but with decidedly mixed results (Chapter 6).

THE 1984 CORONARY PRIMARY PREVENTION TRIAL, A LANDMARK STUDY (CHAPTER 7)

The National Heart, Lung and Blood Institute decided in 1970 to take the bull by the horns and try for the definitive study to validate the lipid hypothesis. Planning for the Coronary Primary Prevention Trial (CPPT) began in 1971 but the results only became available in 1984. A total of 3,806 hypercholesterolemic men without prior coronary heart disease were treated with cholestyramine or with placebo for an average of 7.4 years. LDL cholesterol in the treated group fell by 20.3 percent. The primary end point – definite coronary heart disease death or definite nonfatal myocardial infarction – was 19 percent lower in the cholestyramine group ($p < 0.05$). Secondary end points (new positive exercise test, angina, or coronary bypass surgery) were also reduced and to about the same extent – by 20–25 percent (37;38). With these results added to the increasing body of evidence in favor of the lipid hypothesis, the NIH called a Consensus Conference to evaluate formally the available evidence and to advise what steps if any should be taken. The Expert Panel unanimously concluded that measures should be taken to make testing for and treating hypercholesterolemia best medical practice (39). To that end a National Cholesterol Education Program, coordinated by NIH, was established and charged to offer guidelines to physicians, to health professionals, and to the public on the importance of treating hypercholesterolemia (40;41).

THE VERY VOCAL OPPOSITION

Many if not most investigators warmly endorsed the conclusions of the NIH panel. A lead article in the *Medical Journal of Australia* by Leon A. Simons entitled "The lipid hypothesis is proven" concluded that: "The LRC-CPPT has given a new respectability and credibility to the dietary and pharmacologic management of hypercholesterolemia" (42). *Postgraduate Medicine* put it this way: "Coronary disease prevention: proof of the anticholesterol pudding" (43). Paul Nestel, a leading expert in lipid metabolism, wrote: "Time to treat cholesterol seriously" (44).

On the other hand, the Consensus Conference conclusions were vigorously challenged immediately by a small but vocal group of colleagues. For example, George W. Mann, Associate Professor of Biochemistry at Vanderbilt University College of Medicine, had this to say about the directors of the trial: "They have held repeated press conferences bragging about this cataclysmic break-through which the study directors claim shows that lowering cholesterol lowers the frequency of coronary disease. They have manipulated the data to reach the wrong conclusions." And later: "The managers at NIH have used Madison Avenue hype to sell this failed trial in the way the media people sell an underarm deodorant" (45). Michael Oliver, a highly influential British cardiologist, had this to say: "Those who initiated the idea [of the Consensus Conference] were either naïve or determined to use the forum for special pleading, or both. The panel of jurists ... was selected to include experts who would, predictably, say that ... all levels of blood cholesterol in the United States are too high and should be lowered. Of course, this is exactly what was said" (46). E.H. Ahrens, Jr., whose pioneering work had clearly shown the important impact of diet on blood cholesterol levels wrote an article in the *Lancet* entitled "The diet–heart question in 1985: has it really been settled?" (47). Clearly, he was still not convinced.

This continuing controversy over the years made it very much an uphill battle to convince practitioners, including the cardiologists (perhaps *especially* the cardiologists), to pay attention to hypercholesterolemia. Even with the imprimatur of the NIH behind it and the full-court press to educate both practitioners and the public through the National Cholesterol Education Program (40), most patients with high-risk cholesterol levels were still under-treated if they were treated at all.

While the evidence for the lipid hypothesis became stronger with every passing year, the idea that hypercholesterolemia could be centrally important continued to be rejected, at times quite angrily, by many cardiologists and nutrition experts, especially in the United Kingdom. Asked to reflect back on the attitudes toward the lipid hypothesis in the UK of the 1960s and 1970s, Professor Hugh Tunstall-Pedoe, a distinguished cardiologist, recalled that "Cholesterol was something that Americans had; it definitely wasn't British. Anyone who talked about cholesterol was obviously suffering from American-style hypochondria" (48). In 1976, an editorial by Michael Oliver in the *British Heart Journal* concluded that: "The view that raised plasma cholesterol is *per se* a cause of coronary heart disease is *untenable*" (49). Sir John McMichael, then the pre-eminent British cardiologist, aggressively attacked the hypothesis in a 1979 piece entitled "Fats and atheroma: an inquest" (50). Even E.H. Ahrens, Jr., the man whose own pioneering clinical research showed conclusively that blood cholesterol could indeed be reduced by appropriate changes in diet, took exception to proposals to change the diet of the American public. In 1979, he wrote that such recommendations would be "unwise, impractical, and unlikely to lead to a reduced incidence of arteriosclerotic disease" (51). Michael Oliver was for many years a vocal skeptic regarding the importance of blood cholesterol levels. In 1981, he wrote that "reduction of raised serum cholesterol is a card of uncertain quality in the primary prevention of [coronary heart disease]" and that "reduction of raised serum cholesterol could lead to adverse biological changes" (52). An editorial in the *Journal of the American College of Cardiology*, entitled "The cholesterol pessimist" took issue with Oliver's negative views (53).

In 1973, E.R. Pinckney and C. Pinckney published *The Cholesterol Controversy* (54) and summarized their views on the back cover: "There is no scientific evidence to show that a high blood cholesterol is the cause of heart disease or that lowering cholesterol – if such a thing is possible – will prevent heart troubles of any kind." Frederick J. Stare, Robert E. Olson, and Elizabeth M. Whelan in 1989 subtitled their book "Beyond the Cholesterol Scare" and assured their readers that the campaign to control hypercholesterolemia greatly exaggerated the risk (yet agreeing that coronary heart disease was the major cause of death in the United States) (55). Another persistent critic has been Uffe Ravnskov. Even in the second edition of his book, published in 2000, i.e. after the publication of the two large statin trials (56;57) showing highly significant drops in blood cholesterol and 20–30 percent decreases in both coronary heart disease mortality and total mortality, he continued to dismiss the lipid hypothesis as based on myths (58). In the Appendix are gathered some of the challenges to the lipid hypothesis put forward by these and other critics. The responses to them are briefly summarized there together with cross-references to pages in this book where the issues are dealt with in more detail.

THE BIRTH OF THE STATINS (CHAPTER 8)

The theoretical possibility of lowering blood cholesterol levels by inhibiting endogenous biosynthesis of cholesterol was recognized as early as the 1950s, but the agents tried were ineffective or toxic or both. From the work of Bloch (16) and that of Gould and Popjak (59), it was known that the rate-limiting enzyme in cholesterol biosynthesis was HMGCoA reductase. A drug inhibiting at this site would be ideal, and many attempts were made to find such an inhibitor but without success. In 1976, Akira Endo *et al.*, working at the Sankyo Co. in Tokyo, discovered a powerful competitive inhibitor of HMGCoA reductase in the supernatant of a culture of *Penicillium citrinin* (60). The purified compound, which they designated ML-236B (later called compactin), gave 50 percent inhibition of cholesterol biosynthesis in a cell-free liver preparation at an astonishingly low concentration -2.6×10^{-7} M! Very few drugs are as potent as that. Compactin was the first drug of the statin family and it proved to be both safe and highly effective in lowering cholesterol levels in patients, even those with familial hypercholesterolemia. For reasons still not entirely clear, Sankyo never brought its drug to market. Merck began to screen fungal broths and, in 1980, discovered lovastatin, the first statin to reach the market (61). Now it was possible to drop LDL cholesterol levels by as much as 25–35 percent and the consequent decreases in risk were proportionately greater. There no longer was any doubt about the validity of the lipid hypothesis. Today of course aggressive treatment of hypercholesterolemia has become standard medical practice. This is largely due to the statins. They are remarkably potent, safer than aspirin, easy to take, and easy to prescribe. Even Oliver finally accepted the hypothesis and began actively to proselytize for aggressive treatment (62).

WHERE WE STAND TODAY

When I attended medical school in the 1940s, students were told that atherosclerosis was an inevitable accompaniment of aging. It was a "degenerative disease" about which nothing could be done and that was that. Today, prevention is the order of the day. The disease is eminently treatable and lowering blood cholesterol is one of the most effective ways of doing so. We know that we are winning the war against coronary artery disease (CHD). It can be prevented. In fact, the clinical trials with the statins have shown remarkable decreases in both CHD mortality and also total mortality. Lowering LDL by about 25 percent is enough to lower CHD mortality by 30–40 percent and that is the result of only 5 or 6 years of intervention (56;63–65). It seems reasonable to extrapolate and expect even greater reductions if treatment is started earlier in life and continued, not for just 5 years, but for decades (Chapter 9).

REFERENCES

1. Cornfield, J. 1969. The Bayesian outlook and its application. *Biometrics* 25:617–657.
2. Goodman, S.N. 1999. Toward evidence-based medical statistics. 2: The Bayes factor. *Ann Intern Med* 130:1005–1013.
3. Peters, J.P. and Vanslyke, D.D. 1946. *Quantitative Clinical Chemistry*. Williams and Wilkins, Baltimore.
4. Anitschkow, N.N. and Chalatov, S. 1913. Ueber experimentelle Choleserinsteatose und ihre Bedeutung fur die Entstehung einiger pathologischer Prozesse. *Zentralbl Allg Pathol* 24:1–9.
5. Muller, C. 1939. Angina pectoris in hereditary xanthomatosis. *Arch Int Med* 64:675–700.
6. Keys, A., Aravanis, C., Blackburn, H.W., Van Buchem, F.S., Buzina, R., Djordjevic, B.D., Dontas, A.S., Fidanza, F., Karvonen, M.J., Kimura, N., Lekos, D., Monti, M., Puddu, V., and Taylor, H.L. 1966. Epidemiological studies related to coronary heart disease: characteristics of men aged 40–59 in seven countries. *Acta Med Scand Suppl* 460:1–392.
7. Kannel, W.B., Dawber ,T.R., Kagan, A., Revotskie, N., and Stokes, J., III. 1961. Factors of risk in the development of coronary heart disease – six year follow-up experience. The Framingham Study. *Ann Intern Med* 55:33–50.
8. Kinsell, L.W., Partridge, J., Boling, L., Margen, S., and Michael, G. 1952. Dietary modification of serum cholesterol and phospholipid levels. *J Clin Endocrinol Metab* 12:909–913.
9. Ahrens, E.H., Jr., Blankenhorn, D.H., and Tsaltas, T.T. 1954. Effect on human serum lipids of substituting plant for animal fat in diet. *Proc Soc Exp Biol Med* 86:872–878.
10. Leren, P. 1966. The effect of plasma cholesterol lowering diet in male survivors of myocardial infarction. A controlled clinical trial. *Acta Med Scand Suppl* 466:1–92.
11. Dayton, S., Pearce, M.L., Goldman, H., Harnish, A., Plotkin, D., Shickman, M., Winfield, M., Zager, A., and Dixon, W. 1968. Controlled trial of a diet high in unsaturated fat for prevention of atherosclerotic complications. *Lancet* 2:1060–1062.
12. Miettinen, M., Turpeinen, O., Karvonen, M.J., Elosuo, R., and Paavilainen, E. 1972. Effect of cholesterol-lowering diet on mortality from coronary heart-disease and other causes. A twelve-year clinical trial in men and women. *Lancet* 2:835–838.
13. 1961. Dietary fat and its relation to heart attacks and strokes. Report of the Committee for Medical and Community Program of the American Heart Association. *Circulation* 23:133–136.
14. Steinberg, D. 1970. Progress, prospects and provender. Chairman's address before the Council on Arteriosclerosis, American Heart Association, Dallas, Texas, November 12, 1969. *Circulation* 41:723–728.

15. Frantz, I.D., Jr. and Moore, R.B. 1969. The sterol hypothesis in atherogenesis. *Am J Med* 46:684–690.
16. Bloch, K. 1965. The biological synthesis of cholesterol. *Science* 150:19–28.
17. Gofman, J.W., Lindgren, F.T., and Elliott, H. 1949. Ultracentrifugal studies of lipoproteins of human serum. *J Biol Chem* 179:973–979.
18. Fredrickson, D.S., Levy, R.I., and Lees, R.S. 1967. Fat transport in lipoproteins – an integrated approach to mechanisms and disorders. *N Engl J Med* 276:34–42.
19. Havel, R.J., Eder, H.A., and Bragdon, J.H. 1955. The distribution and chemical composition of ultracentrifugally separated lipoproteins in human serum. *J Clin Invest* 34:1345–1353.
20. Havel, R.J. 1987. Origin, metabolic fate, and metabolic function of plasma lipoproteins. In *Hypercholesterolemia and Atherosclerosis*, D. Steinberg and J.M.Olefsky, eds. Churchill Livingstone, New York:117–142.
21. Mahley, R.W., Innerarity, T.L., Rall, S.C., Jr., and Weisgraber, K.H. 1984. Plasma lipoproteins: apolipoprotein structure and function. *J Lipid Res* 25:1277–1294.
22. Goldstein, J.L., Schrott, H.G., Hazzard, W.R., Bierman, E.L., and Motulsky, A.G. 1973. Hyperlipidemia in coronary heart disease. II. Genetic analysis of lipid levels in 176 families and delineation of a new inherited disorder, combined hyperlipidemia. *J Clin Invest* 52:1544–1568.
23. Ross, R. and Glomset, J.A. 1976. The pathogenesis of atherosclerosis (first of two parts). *N Engl J Med* 295:369–377.
24. Ross, R. and Glomset, J.A. 1976. The pathogenesis of atherosclerosis (second of two parts). *N Engl J Med* 295:420–425.
25. Brown, M.S. and Goldstein, J.L. 1974. Familial hypercholesterolemia: defective binding of lipoproteins to cultured fibroblasts associated with impaired regulation of 3-hydroxy-3-methylglutaryl coenzyme A reductase activity. *Proc Natl Acad Sci U.S.A.* 71:788–792.
26. Brown, M.S. and Goldstein, J.L. 1986. A receptor-mediated pathway for cholesterol homeostasis. *Science* 232:34–47.
27. Brown, M.S., Basu, S.K., Falck, J.R., Ho, Y.K., and Goldstein, J.L. 1980. The scavenger cell pathway for lipoprotein degradation: specificity of the binding site that mediates the uptake of negatively-charged LDL by macrophages. *J Supramol Struct* 13:67–81.
28. Brown, M.S. and Goldstein, J.L. 1983. Lipoprotein metabolism in the macrophage: implications for cholesterol deposition in atherosclerosis. *Annu Rev Biochem* 52:223–261.
29. Fogelman, A.M., Shechter, I., Seager, J., Hokom, M., Child, J.S., and Edwards, P.A. 1980. Malondialdehyde alteration of low density lipoproteins leads to cholesteryl ester accumulation in human monocyte-macrophages. *Proc Natl Acad Sci U.S.A.* 77:2214–2218.
30. Mahley, R.W., Innerarity, T.L., Brown, M.S., Ho, Y.K., and Goldstein, J.L. 1980. Cholesteryl ester synthesis in macrophages: stimulation by beta-very low density lipoproteins from cholesterol-fed animals of several species. *J Lipid Res* 21:970–980.
31. Henriksen, T., Mahoney, E.M., and Steinberg, D. 1981. Enhanced macrophage degradation of low density lipoprotein previously incubated with cultured endothelial cells: recognition by receptors for acetylated low density lipoproteins. *Proc Natl Acad Sci U.S.A.* 78:6499–6503.
32. Kodama, T., Reddy, P., Kishimoto, C., and Krieger, M. 1988. Purification and characterization of a bovine acetyl low density lipoprotein receptor. *Proc Natl Acad Sci U.S.A.* 85:9238–9242.
33. Savenkova, M.L., Mueller, D.M., and Heinecke, J.W. 1994. Tyrosyl radical generated by myeloperoxidase is a physiological catalyst for the initiation of lipid peroxidation in low density lipoprotein. *J Biol Chem* 269:20394–20400.
34. Glass, C.K. and Witztum, J.L. 2001. Atherosclerosis: the road ahead. *Cell* 104:503–516.
35. Hansson, G.K. 2001. Immune mechanisms in atherosclerosis. *Arterioscler Thromb Vasc Biol* 21:1876–1890.
36. Libby, P., Ridker, P.M., and Maseri, A. 2002. Inflammation and atherosclerosis. *Circulation* 105:1135–1143.

37. 1984. The Lipid Research Clinics Coronary Primary Prevention Trial results. I. Reduction in incidence of coronary heart disease. *JAMA* 251:351–364.

38. 1984. The Lipid Research Clinics Coronary Primary Prevention Trial results. II. The relationship of reduction in incidence of coronary heart disease to cholesterol lowering. *JAMA* 251:365–374.

39. 1985. Consensus conference. Lowering blood cholesterol to prevent heart disease. *JAMA* 253:2080–2086.

40. Cleeman, J.I. 1989. National Cholesterol Education Program. Overview and educational activities. *J Reprod Med* 34:716–728.

41. 1988. Report of the National Cholesterol Education Program Expert Panel on detection, evaluation, and treatment of high blood cholesterol in adults. The Expert Panel. *Arch Intern Med* 148:36–69.

42. Simons, L.A. 1984. The lipid hypothesis is proven. *Med J Aust* 140:316–317.

43. Podell, R.N. 1984. Coronary disease prevention: proof of the anticholesterol pudding. *Postgrad Med* 75:193–196.

44. Nestel, P.J. 1984. Time to treat cholesterol seriously. *Aust N.Z. J Med* 14:198–199.

45. Mann, G.V. 1985. Coronary heart disease – "doing the wrong thing." *Nutrition Today* 12–14.

46. Oliver, M.F. 1985. Consensus or nonsensus conferences on coronary heart disease. *Lancet* 1:1087–1089.

47. Ahrens, E.H. 1985. The diet–heart question in 1985 – has it really been settled? *Lancet* 1:1085–1087.

48. 2006. Cholesterol, atherosclerosis and coronary disease in the UK, 1950–2000. *Wellcome Witnesses to Twentieth Century Medicine* 27.

49. Oliver, M. 1976. Dietary cholesterol, plasma cholesterol and coronary heart disease. *Br Heart J* 38: 214–218.

50. McMichael, J. 1979. Fats and atheroma – an inquest. *Br Med J* 1:173–175.

51. Ahrens, E.H. 1979. Dietary fats and coronary heart disease: unfinished business. *Lancet* 2:1345–1348.

52. Oliver, M.F. 1981. Serum-cholesterol the knave of hearts and the joker. *Lancet* 2:1090–1095.

53. Henry, P.D. 1988. The cholesterol pessimist. *J Am Coll Cardiol* 12:818–819.

54. Pinckney, E.R. and Pinckney, C. 1973. *The Cholesterol Controversy*. Sherburne Press, Los Angeles:1–162.

55. Stare, F.J., Olson, R.E., and White, A.L. 1989. *Balanced Nutrition: Beyond the Cholesterol Scare*. Bob Adams, Inc., Holbrook, MA:1–360.

56. Scandinavian Simvistatin Survival Study Group. 1994. Randomised trial of cholesterol lowering in 4444 patients with coronary heart disease: the Scandinavian Simvistatin Survival Study (4S). *Lancet* 344:1383–1389.

57. Shepherd, J., Cobbe, S.M., Ford, I., Isles, C.G., Lorimer, A.R., Macfarlane, P.W., McKillop, J.H., and Packard, C.J. 1995. Prevention of coronary heart disease with pravastatin in men with hypercholesterolemia. West of Scotland Coronary Prevention Study Group. *N Engl J Med* 333:1301–1307.

58. Ravnskov, U. 2000. *The Cholesterol Myths*. New Trends Publishing, Inc., Washington, DC: 1–297.

59. Gould, R.G. and Popjak, G. 1957. Biosynthesis of cholesterol in vivo and in vitro from DL-beta-hydroxy-beta-methyl-delta [2-14C]-valerolactone. *Biochem J* 66:51P.

60. Endo, A., Kuroda, M., and Tsujita, Y. 1976. ML-236A, ML-236B, and ML-236C, new inhibitors of cholesterogenesis produced by Penicillium citrinium. *J Antibiot (Tokyo)* 29:1346–1348.

61. Alberts, A.W., Chen, J., Kuron, G., Hunt, V., Huff, J., Hoffman, C., Rothrock, J., Lopez, M., Joshua, H., Harris, E., Patchett, A., Monaghan, R., Currie, S., Stapley, E., Albers-Schomberg, G.,

Hensens, O., Hirshfield, J., Hoogstein, K., Liesch, J., and Springer, J. 1980. Mevinolin: a highly potent competitive inhibitor of hydroxymethylglutaryl-coenzyme A reductase and a cholesterol-lowering agent. *Proc Natl Acad Sci U.S.A.* 77:3957–3961.

62. Oliver, M.F. 1995. Statins prevent coronary heart disease. *Lancet* 346:1378–1379.

63. 2002. MRC/BHF Heart Protection Study of cholesterol lowering with simvastatin in 20,536 high-risk individuals: a randomised placebo-controlled trial. *Lancet* 360:7–22.

64. Law, M.R., Wald, N.J., and Rudnicka, A.R. 2003. Quantifying effect of statins on low density lipoprotein cholesterol, ischaemic heart disease, and stroke: systematic review and meta-analysis. *Br Med J* 326:1423.

65. Tobert, J.A. 1996. Statin therapy and CHD. *Lancet* 347:128.

CHAPTER 2

Animal Models of Atherosclerosis

NIKOLAI N. ANITSCHKOW[1] AND THE CHOLESTEROL-FED RABBIT

Many attempts were made in the late 19th and early 20th centuries to mimic human atherosclerosis in an animal model, including injecting adrenalin, injecting bacteria or bacterial products, and directly traumatizing the arteries in various ways (1). While these interventions certainly damaged the arteries, the lesions were simply not like those found in human atherosclerosis. Then, in 1913, Anitschkow and Chalatov showed that just feeding rabbits purified cholesterol dissolved in pure sunflower oil induced vascular lesions closely resembling those of human atherosclerosis, both grossly and microscopically (2;3). Anitschkow at the time was a young student at the Military Medical Academy in St. Petersburg, working under the mentorship of L. Sobolev, who had assigned him and Chalatov to follow up on earlier studies by Ignatowski (4), studies showing that lesions very much like those in human atherosclerosis could be produced by feeding rabbits diets rich in meat, eggs, and milk. Anitschkow and Chalatov confirmed Ignatowski's findings and set out to determine which dietary component or components were specifically responsible. Feeding just eggs worked. Feeding just the egg yolks worked. Finally, feeding just pure cholesterol purified from the yolks and dissolved in sunflower oil worked. Controls fed only the sunflower oil showed no lesions.

It is fair to say that this paper marked the beginning of the modern era of atherosclerosis research. Over the next few years Anitschkow and his colleagues (1) established:

1. That in the earliest lesions, the fatty streaks, most of the lipid was found in cells containing large numbers of lipid-containing vacuoles (foam cells) (Figure 2.1). These were stained by lipophilic dyes and contained

The Cholesterol Wars by D. Steinberg.
Copyright © 2007 Elsevier Inc. All rights of reproduction in any form reserved.

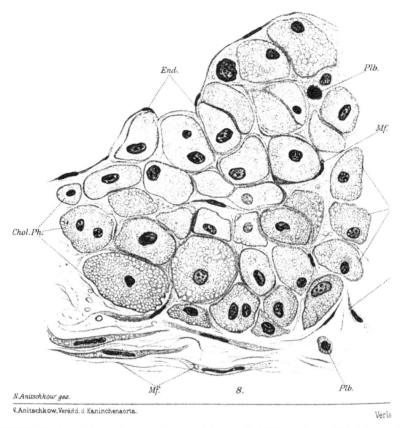

FIGURE 2.1 Anitschkow's drawing of a typical foam cell-rich lesion in a rabbit fed a total of 82.7 g of pure cholesterol in sunflower oil over a period of 139 days. Source: (3).

birefringent droplets (liquid crystals of cholesterol esters). As shown in Figure 2.1, the monolayer of endothelial cells over the lesion appears to be intact.

2. That the very earliest lesions appeared at the root of the aorta and in the aortic arch and then proceeded caudally (Figure 2.2).

3. That there was a characteristic pattern of distribution of early lesions (Figure 2.3). They were most severe at arterial branch points and Anitschkow correctly surmised that this localization was most likely determined by hemodynamic factors.

4. That over long periods of cholesterol feeding there was ultimately deposition of connective tissue (conversion of the fatty streak to the fibrous plaque) and development of a fibrous cap (Figure 2.4).

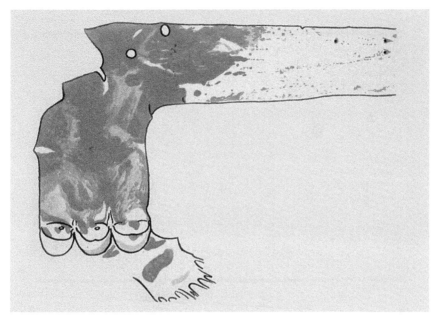

FIGURE 2.2   Aorta of a rabbit fed 61 egg yolks over a 70-day period and stained with the lipophilic dye, Sudan IV. Anitschkow recognized that the earliest lesions appeared in the arch near the orifices of branch points and then moved caudally. Source: (1). This figure is reproduced in color in the color plate section.

5. That early lesions were partially reversible but that the reversal was a very slow process. Not all of the lipids could be mobilized from advanced lesions, leaving behind the fibrous cap and a few cholesterol crystals (Figure 2.5).

6. That the extent of lesions was proportional to the degree of blood cholesterol elevation and the duration of exposure to it. This is a crucially important point. Anitschkow was well aware that it was the level of *blood* cholesterol reached that determined the size and extent of lesions. He pointed out that the lipid deposits only occurred in animals fed cholesterol and oil for a considerable length of time. He wrote:

> The blood of such animals exhibits an enormous increase in cholesterin [cholesterol] content, which in some cases amounts to several times the normal quantity. It may therefore be regarded as certain that in these experimental animals large quantities of the ingested cholesterin [cholesterol] are absorbed, and that the accumulations of this substance in the tissues can only be interpreted as deposits of lipoids circulating in large quantities in the humors of the body.

7. That the cholesterol-loaded cells were probably white blood cells that had infiltrated the artery wall. Thus he anticipated that inflammation might play a role in lesion development.

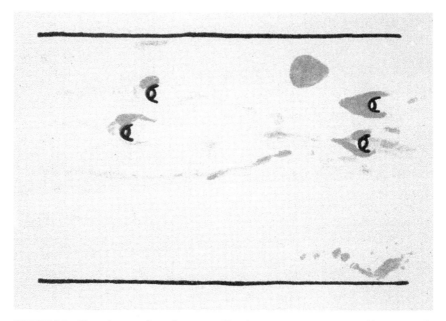

FIGURE 2.3   Thoracic aorta from the same rabbit shown in Figure 2.2. Anitschkow minutely
described the appearance of "small yellowish spots of a triangular or semilunar shape, situated close
below the orifices of [intercostal arteries]." He noted, in the abdominal aorta, "spur shaped thickenings ... below the orifices, which probably serve the formation or regulation of various currents in
the blood stream ... which are "... principally affected by the atherosclerotic changes." He clearly
stated the hemodynamic hypothesis for specific localization of lesions. Source: (1). This figure is
reproduced in color in the color plate section.

One of the things that distinguished Anitschkow's model from many others
that had been proposed is that it required no direct injury or damage to the
arteries. The possibility that atherosclerotic lesions might be the result of some
form of injury was a hypothesis widely tested during the early 20th century. In
his 1933 review, Anitschkow lists more than a dozen papers describing efforts
to induce atherosclerotic lesions using mechanical injury (ligation, pulling,
pinching, wounding, cauterization with silver nitrate or galvanic wire) (1). He
sums up the results this way: "these different ways of causing lesions in the arteries resulted merely in the production of inflammatory arterial changes which had
no similarity with human arteriosclerosis." He goes on to propose that what such
injuries do is to alter the wall of the artery in such a way as to make it more susceptible to lesion formation *if there is concomitant hypercholesterolemia*. He cites
the work of Ssolowjew who showed that injury such as cauterization leads to
"regenerative thickening of the intima" but no lipid deposition. In contrast, if
cholesterol feeding is started *before* the injury is produced, the site of the injury
shows more severe atherosclerotic changes and greater lipid deposition (5).

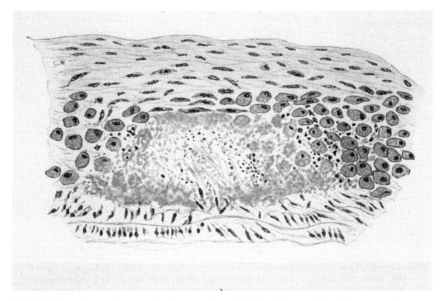

FIGURE 2.4   Advanced plaque in the aorta of a rabbit fed cholesterol for 124 days and then put back on a chow diet for 101 days before sacrifice. Anitschkow calls attention to the central necrotic lipid core containing needle-like crystals of cholesterol, scattered calcium granules, groups of foam cells on either side of the core, and a fibrous cap overlying the lesion. Source: (1). This figure is reproduced in color in the color plate section.

Anitschkow's conclusion was that "lesions of the arterial wall which are associated with hyperplasia of the intima create a local predisposition to the formation of lipoidal deposits, that is to the development of atherosclerosis, *provided that they are synchronous with hypercholesterolemia*" (emphasis added). This prescient conclusion has been borne out by many later studies differentiating lesions due to arterial injury without hypercholesterolemia from those produced in the presence of hypercholesterolemia.

Anitschkow's dictum "No atherosclerosis without cholesterol" has often been cited as evidence that he was unaware of the multifactorial nature of the disease. However, his 1933 review gives the lie to this when he sums up as follows: "The views here set forth concerning the etiology of atherosclerosis constitute what I have called the 'combination theory' of its origin." He was fully aware that the degree of atherosclerosis, while perhaps most evidently dependent on the degree of blood cholesterol elevation, could be significantly affected by blood pressure, toxic substances, local arterial changes, and the like.

Anitschkow was a keen-eyed structural pathologist and a careful experimentalist; he thought in terms of function and time-related pathogenesis. If the full significance of his findings had been appreciated at the time, we might

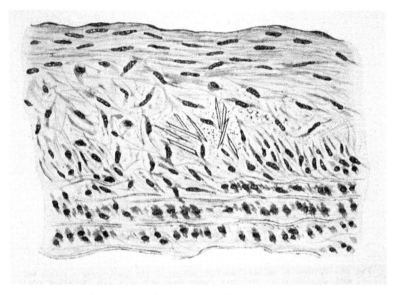

FIGURE 2.5 Atherosclerotic plaque from the aorta of a rabbit fed cholesterol for 106 days, i.e. about the same as the rabbit shown in Figure 2.4, but then returned to a chow diet for 785 days – over 2 years! – before sacrifice. Anitschkow describes it as follows: "The lipoidal masses have disappeared; only a few cholesterin [cholesterol] crystals and lipoid-containing wandering-cells are still present." He adds that the surface is "fibrous and dense." He recognized that even late lesions were at least partially reversible, i.e. at least the lipid from them. Source: (1). This figure is reproduced in color in the color plate section.

have saved more than 30 years in the long struggle to settle the cholesterol controversy and Anitschkow might have won a Nobel Prize (Figure 2.6) (6).

Instead, his findings were largely rejected or at least not followed up. Serious research on the role of cholesterol in human atherosclerosis did not really get underway until the 1940s. Why?

WHY WASN'T ANITSCHKOW'S LEAD FOLLOWED UP?

Some laboratories did indeed try to reproduce Anitschkow's findings. For example, Bailey at Stanford, using rabbits and guinea pigs, quickly confirmed Anitschkow (7;8). Most investigators, however, instead of using rabbits, used the laboratory animals they were more familiar with – rats or dogs. Cholesterol feeding in these species failed to induce lesions. Understandably, these investigators concluded that Anitschkow's results must reflect some peculiarity of the rabbit. After all, they said, it is a strict herbivore that normally has zero cholesterol intake and only a very low fat intake. The rabbit model was dismissed as irrelevant to the human disease.[2]

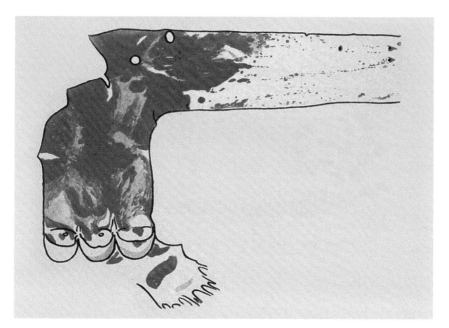

FIGURE 2.2 Aorta of a rabbit fed 61 egg yolks over a 70-day period and stained with the lipophilic dye, Sudan IV. Anitschkow recognized that the earliest lesions appeared in the arch near the orifices of branch points and then moved caudally. Source: (1).

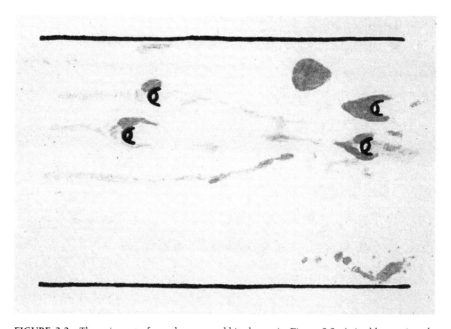

FIGURE 2.3 Thoracic aorta from the same rabbit shown in Figure 2.2. Anitschkow minutely described the appearance of "small yellowish spots of a triangular or semilunar shape, situated close below the orifices of [intercostal arteries]." He noted, in the abdominal aorta, "spur shaped thickenings ... below the orifices, which probably serve the formation or regulation of various currents in the blood stream ... which are "... principally affected by the atherosclerotic changes." He clearly stated the hemodynamic hypothesis for specific localization of lesions. Source: (1).

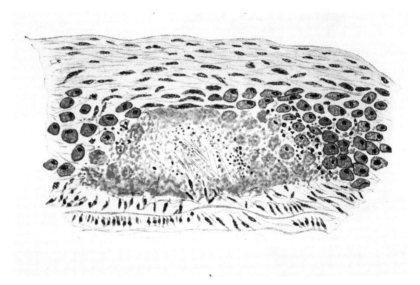

FIGURE 2.4 Advanced plaque in the aorta of a rabbit fed cholesterol for 124 days and then put back on a chow diet for 101 days before sacrifice. Anitschkow calls attention to the central necrotic lipid core containing needle-like crystals of cholesterol, scattered calcium granules, groups of foam cells on either side of the core, and a fibrous cap overlying the lesion. Source: (1).

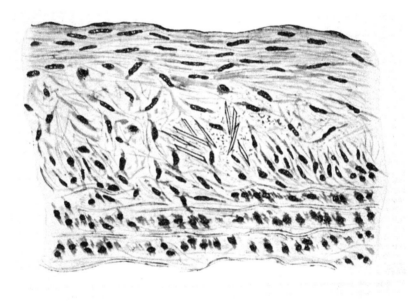

FIGURE 2.5 Atherosclerotic plaque from the aorta of a rabbit fed cholesterol for 106 days, i.e. about the same as the rabbit shown in Figure 2.4, but then returned to a chow diet for 785 days – over 2 years! – before sacrifice. Anitschkow describes it as follows: "The lipoidal masses have disappeared; only a few cholesterin [cholesterol] crystals and lipoid-containing wandering-cells are still present." He adds that the surface is "fibrous and dense." He recognized that even late lesions were at least partially reversible, i.e. at least the lipid from them. Source: (1).

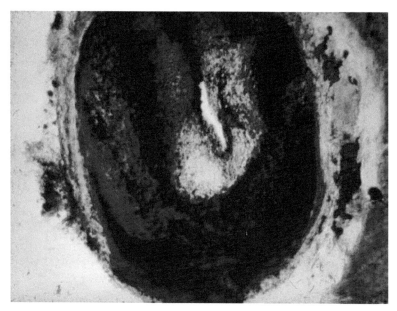

FIGURE 5.7 (a). Almost totally obstructed coronary artery of a Rhesus monkey fed an atherogenic diet by Armstrong *et al.* for 17 months. Blood cholesterol level was about 700 mg/dl.

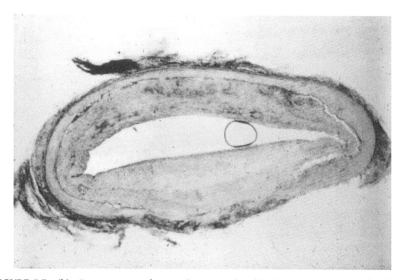

FIGURE 5.7 (b). Coronary artery from a Rhesus monkey fed the same atherogenic diet for 17 months but then switched back to a low-fat, low-cholesterol diet for 40 months before sacrifice. Almost all of the stainable lipid has disappeared, the lumen is restored almost to normal and the artery has remodeled considerably. A thick fibrous coat remains. See text for further details of the protocol. Source: photomicrographs courtesy of Dr. William E. Connor.

Nikolai Nikolaevich Anitschkow (In
Commemoration of the 90th Anniversary of his
Birthday) (1975)

N. N. Anitschkow

FIGURE 2.6   Nikolai Nikolaevich Anitschkow (1985–1964) shortly before his retirement from
the Institute for Experimental Medicine in St. Petersburg. Source: (6).

What was not appreciated in Anitschkow's day was the fact that rats and
dogs, unlike rabbits, are very efficient in converting cholesterol to bile acids.
Consequently, even on very high intakes of dietary cholesterol the *blood* cholesterol in these species does not rise appreciably. Steiner and Kendall, 33 years later,
would show that inhibiting thyroid function in dogs and *then* feeding them cholesterol does increase blood cholesterol, and *then does induce lesions* (11). So here was
one reason Anitschkow's work was not taken seriously: failure to recognize the
three-step nature of what was going on – feeding of cholesterol, followed by elevation of blood cholesterol levels, followed by atherogenesis. Only if the second step
kicks in do you get atherosclerosis. Actually Anitschkow had himself tried to
induce lesions in dogs and noted that they did not respond. He speculated that the
dog, a carnivore, was adapted to eating fat-rich and cholesterol-rich foods and
could therefore excrete excess cholesterol, a remarkable metabolic insight for the
early 20th century!

Another reason Anitschkow's findings were not taken seriously is that the blood cholesterol levels in his early studies were extraordinarily high – 500–1,000 mg/dL or even higher. The argument was that human levels were almost never that high and that extrapolation was unwarranted. This was a legitimate reservation at the time but soon after his original studies Anitschkow showed that more modest elevations of cholesterol levels in rabbits were sufficient to induce lesions. It just took longer.

Were his findings not widely known? Was that the reason they were not followed up more aggressively? Not at all. He published in German in the most respected and widely read journals of the time and he published a long series of papers on the subject over the next few years. As mentioned above, a number of laboratories did try to reproduce his results, so we know that many investigators were aware of his work. In 1933, Anitschkow published (in English) an extensive review of the work of his laboratory and that of others in *Arteriosclerosis*, a widely quoted collection of authoritative reviews edited by Cowdry (1). So at least the community of scholars interested in the pathogenesis of atherosclerosis was aware, or at least should have been aware, of his work.

There is another and possibly more important reason for the indifferent response of the scientific community. Anitschkow's findings ran counter to the prevailing view of atherosclerosis. Atherosclerosis was generally accepted to be an inevitable accompaniment of aging (the "Senescence Hypothesis"): it was a chronic, slowly progressive deterioration developing over decades. How could one possibly expect to mimic such a disease – the argument went – by feeding cholesterol to young rabbits for just weeks or months? It seemed totally implausible. There were lesions to be sure, but they could hardly be considered a model for human atherosclerosis (even though Anitschkow himself insisted they looked very much like human lesions).

In retrospect, Anitschkow's body of work showed clearly and convincingly that hypercholesterolemia in rabbits was a sufficient cause of atherosclerosis. Of course it did not necessarily follow that cholesterol – either in the diet or in the blood – was also an important factor in *human* atherosclerosis. That conclusion would have to await studies showing that hypercholesterolemia in humans was indeed associated with atherosclerosis and, ultimately, clinical trials to establish that relationship as a causal one. However, Anitschkow's work should have galvanized the scientific community and started a more serious approach to this major human disease problem many years ago.

Here was a classic example of how rigid, preconceived ideas can stand in the way of scientific progress. An opportunity was lost. Why did no one ask the (now) obvious questions: How is the cholesterol carried in the rabbit blood? How does it get into the arterial wall? Which white blood cells are entering the artery wall and taking up huge amounts of cholesterol? Does the diet, especially the fat and cholesterol in it, increase blood cholesterol in humans? Answers would come, but only about 40 years later. The main point is that those questions were apparently not even asked at the time.

Anitschkow's work was done in St. Petersburg/Leningrad between 1912 and 1964, first in the Military Medical Academy and later at the Institute of Experimental Medicine. In 1962, just 2 years before his death, Anitschkow established a new Laboratory of Lipid Metabolism within the Institute and named Dr. Anatoly N. Klimov to head it. A few years later Klimov and collaborators carried out an experiment of heroic proportions to show definitively that the atherosclerosis in Anitschkow's rabbits was a direct result of the elevated plasma lipoproteins and not some other indirect consequence of the cholesterol-rich diet. They isolated serum from cholesterol-fed rabbits, removed the chylomicrons by centrifugation and gave the hyperlipidemic serum intravenously to chow-fed recipients. Over a 5–7-month period the recipients received 14–25 g of cholesterol intravenously and developed significant arterial lesions. So blood cholesterol, mainly in LDL and VLDL (very low density lipoproteins), was the immediate and sufficient cause of the atherosclerosis (12). There is something satisfying in the continuity of focus on atherosclerosis in this Institute, where Klimov remained as head of the Department of Biochemistry until his retirement.

WHAT LED ANITSCHKOW TO FEED RABBITS CHOLESTEROL?

Before leaving the Anitschkow story, it is of interest to ask just what inspired those classical studies. Like so many breakthroughs in science, it seems that Anitschkow's discovery was serendipitous. Many have probably assumed that Anitschkow was led to do his experiment by the Windaus 1910 paper reporting that the aortas of patients with atherosclerosis contained much higher concentrations of cholesterol than did normal aortas (13). Actually, the rationale for Anitschkow's studies, as pointed out by Hoeg and Klimov (14), actually came from a 1909 paper by Ignatowski, another young experimental pathologist working at the Military Medical Academy in St. Petersburg (4). Ignatowski was pursuing a hypothesis put forward some years earlier by Nobel Prize-winning microbiologist, I. Metschnikow. Metschnikow proposed that an excess of protein in the diet was potentially toxic and somehow accelerated the aging process. So Ignatowski decided to feed rabbits a protein-rich diet and look for signs of such toxicity. He fed his rabbits large amounts of meat, eggs, and milk. These protein-rich diets were indeed toxic in young rabbits, affecting liver and adrenals, but in adult rabbits the major effect was the development of striking arterial lesions resembling those of human atherosclerosis (4). Since atherosclerosis was considered one of the hallmarks of aging, Ignatowski considered that his findings had confirmed Metschnikow's "protein toxicity" hypothesis. One of Ignatowski's fellows, N.W. Stuckey, extended these studies and showed that the protein in the diet used by Ignatowski was actually not necessary (15). Egg yolk alone gave the same results, as did feeding of beef brain. The lipid, not the protein, was responsible. It remained for Anitschkow and Chalatov progressively to narrow things down and show that the vascular damage could be induced simply by feeding pure cholesterol derived from egg yolks – without the eggs or milk or meat, i.e. without the protein or other lipids. Another instance of an unpleasant fact destroying a beautiful hypothesis. However, in this instance the new findings generated a very valuable new hypothesis, the lipid hypothesis of atherogenesis.

ATHEROSCLEROSIS IN OTHER SPECIES

As mentioned above, some species (e.g. rat and dog) seem to be resistant to cholesterol-induced atherosclerosis. However, that is not because their arteries are immune but because their blood cholesterol does not go up very much on the fat-rich, cholesterol-rich diet that works nicely in rabbits, guinea pigs, and goats. If the investigator can find a way to raise the blood cholesterol sufficiently, then the dog (11) and the rat (16) do indeed develop typical lesions. A list of species in which experimental atherosclerosis has been produced is shown in Table 2.1. Perhaps a few have been inadvertently omitted. The point is

TABLE 2.1   Animal species in which experimental atherosclerosis has been successfully induced by raising blood cholesterol levels

Species	Reference
Baboon	(19)
Cat	(20)
Chicken	(21;22)
Chimpanzee	(19)
Dog	(11)
Goat	(23)
Guinea pig	(24)
Hamster	(25)
Monkey	(19)
Mouse	(17;18)
Parrot	(21)
Pig	(26)
Pigeon	(27)
Rabbit	(2)
Rat	(28;29)

very clearly made: the arteries of virtually every animal species are susceptible to this disease if only the blood cholesterol level can be raised enough and maintained high enough for a long enough period of time. The mouse and the rat were at one time thought to be totally resistant but then in 1958 Wilgram and co-workers showed that on the proper regimen, albeit a rather drastic one, rats developed extensive lesions (16). In 1985 Paigen identified a mouse strain (C57BL/6J) that developed lesions, albeit modest in extent, when fed a diet rich in cholesterol, saturated fat and cholic acid (17). In 1992 Plump et al., in the laboratory of Jan Breslow at Rockefeller University, described the dramatic, extensive and human-like lesions produced by knocking out the apoprotein E gene in the mouse (18). That work opened up a new era in atherosclerosis research. Now it was possible to critically and much more easily test postulated pathogenetic factors using mouse gene manipulation. Today the most studied animal model is the mouse model, once thought to be as irrelevant as Anitschkow's rabbits.

NOTES

1   There is no accepted convention for transliterating Russian names. I chose "Anitschkow" because he himself spelled it that way when he published in English and that is the spelling used by Pub Med. Other spellings have included "Anitchkov," "Anitchkow," "Anichkov," "Anichkow," and perhaps others.
2   Many years later Yoshiio Watanabe in Japan would discover a naturally occurring but rare mutation in rabbits that mimics exactly one of the mutations that causes familial hypercholesterolemia in humans. These rabbits develop blood cholesterol levels over 600 mg/dl *on their regular chow diet*. No added dietary fat or cholesterol is necessary. And they develop atherosclerosis just as severe as that in the wild-type cholesterol-fed rabbits (9;10).

REFERENCES

1. Anitschkow, N. 1933. Experimental atherosclerosis in animals. In *Arteriosclerosis*, E.V. Cowdry, ed. Macmillan, New York:271–322.
2. Anitschkow, N.N. and Chalatov, S. 1913. Ueber experimentelle Choleserinsteatose und ihre Bedeutung fur die Entstehung einiger pathologischer Prozesse. *Zentralbl Allg Pathol* 24:1–9.
3. Anitschkow, N. 1913. Ueber die Veranderungen der Kaninchenaorta bei experimenteller Cholesterinsteatose. *Beitr Pathol Anat* 56:379–404.
4. Ignatowski, A. 1909. Uber die Wirkung des Tierischen Eiweisses auf die Aorta und die paerenchymatosen Organe der Kaninchen. *Virchows Arch fur path Anat* 198:248–270.
5. Ssolowjew, A. 1929. Experimentelle Untersuchungen uber die Bedeutung von lokaler Schadigung fur die Lipoidablagerung in der Arterienwand. *Zeitschrift fur die gesamte experimentelle Medizin* 69:94–104.
6. Khavkin, T.N. and Pozhariski, K.M. 1975. Nikolai Nikolaevich Anitschkow. *Beitr z Path* 156:301–312.
7. Bailey, C.H. 1915. Observations on cholesterol-fed guinea pigs. *Proc Soc Exper Biol* 13:60–62.
8. Bailey, C.H. 1916. Atheroma and other lesions produced in rabbits by cholesterol feeding. *J Exp Med* 23:69–84.

9. Watanabe, Y. 1980. Serial inbreeding of rabbits with hereditary hyperlipidemia (WHHL-rabbit). *Atherosclerosis* 36:261–268.

10. Kita, T., Brown, M.S., Watanabe, Y., and Goldstein, J.L. 1981. Deficiency of low density lipoprotein receptors in liver and adrenal gland of the WHHL rabbit, an animal model of familial hypercholesterolemia. *Proc Natl Acad Sci U.S.A.* 78:2268–2272.

11. Steiner, A. and Kendall, F.E. 1946. Atherosclerosis and arteriosclerosis in dogs following ingestion of cholesterol and thiouracil. *Arch Path* 42:433–444.

12. Klimov, A.N., Rodionova, L.P., and Petrova-Maslakova, L.G. 1966. Experimental atherosclerosis induced by repeated intravenous administration of hypercholesterolaemic serum. *Cor Vasa* 8:225–230.

13. Windaus, A. 1910. Uber den Gehalt nirmaler und atheromatoser Aorten an Cholsterin und Cholesterinestern. *Hoppe-Seyler Z Physiol Chemie* 67:174–176.

14. Hoeg, J.M. and Klimov, A.N. 1993. Cholesterol and atherosclerosis: "the new is the old rediscovered." *Am J Cardiol* 72:1071–1072.

15. Stuckey, N.W. 1912. Ueber die Veranderungene der Kaninchenaorta bei der Futterung mit verschiedenen Fettsorten. *Centralblatt fur Allgemeine Pathologie und Pathologische Anatomie* 23:910–911.

16. Wilgram, G.F. 1958. Dietary method for induction of atherosclerosis, coronary occlusion and myocardial infarcts in rats. *Proc Soc Exp Biol Med* 99:496–499.

17. Paigen, B., Morrow, A., Brandon, C., Mitchell, D., and Holmes, P. 1985. Variation in susceptibility to atherosclerosis among inbred strains of mice. *Atherosclerosis* 57:65–73.

18. Plump, A.S., Smith, J.D., Hayek, T., Aalto-Setala, K., Walsh, A., Verstuyft, J.G., Rubin, E.M., and Breslow, J.L. 1992. Severe hypercholesterolemia and atherosclerosis in apolipoprotein E-deficient mice created by homologous recombination in ES cells. *Cell* 71:343–353.

19. Blaton, V. and Peeters, H. 1976. The nonhuman primates as models for studying human atherosclerosis: studies on the chimpanzee, the baboon and the rhesus macacus. *Adv Exp Med Biol* 67:33–64.

20. Ginzinger, D.G., Wilson, J.E., Redenbach, D., Lewis, M.E., Clee, S.M., Excoffon, K.J., Rogers, Q.R., Hayden, M.R., and McManus, B.M. 1997. Diet-induced atherosclerosis in the domestic cat. *Lab Invest* 77:409–419.

21. Anitschkow, N. 1925. Einige Ergebnisse der experimentellen Atherosklerosforschung. *Verhandlungen den Deutschen Pathologischen Gesellschaft* 20:149–154.

22. Katz, L.N. and Pick, R. 1961. Experimental atherosclerosis as observed in the chicken. *J Atheroscler Res* 1:93–100.

23. Chalatow, S. 1929. Bemerkungen an den Arbeiten uber Cholesterinsteatose. *Virchow's Archiv f path Anat* 272:691–708.

24. Anitschkow, N. 1922. Ueber die experimentelle Atherosklerose der Aorta beim Meerschwinchen. *Beitr Pathol Anat* 70:265–281.

25. Nistor, A., Bulla, A., Filip, D.A., and Radu, A. 1987. The hyperlipidemic hamster as a model of experimental atherosclerosis. *Atherosclerosis* 68:159–173.

26. Rowsell, H.C., Downie, H.G., and Mustard, J.F. 1958. The experimental production of atherosclerosis in swine following the feeding of butter and margarine. *Can Med Assoc J* 79:647–654.

27. St Clair, R.W., Toma, J.J., Jr., and Lofland, H.B. 1974. Chemical composition of atherosclerotic lesions of aortas from pigeons with naturally occurring or cholesterol-aggravated atherosclerosis. *Proc Soc Exp Biol Med* 146:1–7.

28. Wissler, R.W., Eilert, M.L., Schroeder, M.A., and Cohen, L. 1954. Production of lipomatous and atheromatous arterial lesions in the albino rat. *AMA Arch Pathol* 57:333–351.

29. Andrus, S.B., Fillios, L.C., Mann, G.V., and Stare, F.J. 1956. Experimental production of gross atherosclerosis in the rat. *J Exp Med* 104:539–554.

CHAPTER 3

Hypercholesterolemia and Atherosclerosis in Humans: Causally Related?

THE CLINICAL AND GENETIC EVIDENCE

Probably the first hints that human coronary heart disease might be linked to cholesterol or other lipids came from scattered anecdotal case reports of children with xanthomas – large deposits of lipids attached to tendon sheaths on the backs of the hands or at the ankles, or just beneath the skin on elbows, knees, or buttocks. These deposits are benign but sometimes unacceptably disfiguring, so the parents consult their family physician or a dermatologist. A number of these children developed serious heart problems at a startlingly early age. In 1889, Lehzen and Knauss (1) reported the case of a child that had had xanthomatosis since age 3 and who died suddenly at age 11. Postmortem examination revealed extensive xanthomatous deposits in the aorta and other large arteries, including the coronary arteries. Her sister, age 9, also had cutaneous xanthomas. In retrospect this was clearly a case of homozygous familial hypercholesterolemia. Of course such cases are extremely rare (about one per million births) and the linkage between xanthomas and coronary arteries remained unclear. Much more common were xanthomas associated with liver disease or diabetes or, most common of all, xanthomas limited to the eyelids (xanthelasma), which can occur in some healthy older people. Some pathologists, while recognizing the presence of lipids in such lesions, nevertheless maintained that they were basically benign connective tissue tumors, akin to fibromas or sarcomas. Others, noting the presence of hypercholesterolemia in some cases, concluded instead that they were not true tumors but rather the result of deposition of cholesterol from the blood. Pinkus and Pick were the first to describe the occurrence of an increase in blood cholesterol and of doubly refractile lipids in the lesions, showing that the stored lipid was not triglyceride but rather cholesterol esters (2). They correctly inferred that the cholesterol was being deposited from the blood into the tendons and into the vascular wall.

The Cholesterol Wars by D. Steinberg.
Copyright © 2007 Elsevier Inc. All rights of reproduction in any form reserved.

Anitschkow, while working in Aschoff's laboratory in Germany, showed that cholesterol-fed rabbits formed xanthoma-like lesions if subcutaneous inflammation was artificially induced. The cells involved were postulated to be derived from the reticuloendothelial system, which Aschoff had only recently described. Similar conclusions were reached about human xanthomas by several authors over the next two or three decades but the evidence remained limited and anecdotal (3–6).

Siegfried J. Thannhauser, internationally known for his studies of the lipidoses and his classification of xanthomatous diseases, was well aware of the occasional concurrence of hypercholesterolemia and vascular disease with xanthomatosis. He recognized that these cases were quite distinct and he designated a separate category for what he called *primary essential xanthomatosis of the hypercholesterolemic type*. However, in his comprehensive 1938 review of xanthomatous diseases (7), he rejected the idea that the cholesterol ester in the lesions was deposited there from the blood. Instead he concluded that primary essential xanthomatosis of the hypercholesterolemic type was analogous to Gaucher disease and Niemann-Pick disease, lipid storage diseases characterized, respectively, by accumulation of cerebrosides and sphingomyelin. His view at the time was that the accumulation of cholesterol was due to a local metabolic disturbance *in the cells*. In neither Gaucher disease nor Niemann-Pick disease is there an increase in the blood levels of the lipid being stored, indicating that the storage is probably not due to deposition at the involved sites. Seeking a unifying hypothesis, Thannhauser chose to consider the hypercholesterolemia in essential xanthomatosis as "an exception that proves the rule." Apparently he assumed that the hypercholesterolemia was secondary to release of cholesterol from the xanthomas into the blood rather than the driving force causing the lesions.

In 1939, Carl Müller (Figure 3.1), a Norwegian professor of internal medicine, published a now-classic paper in which he reviewed the already significant literature on the concurrent familial expression of xanthomatosis, hypercholesterolemia and heart disease and added observations on 76 cases from 17 Norwegian families (8).

Consanguineous marriages in some isolated communities in Norway were at the time still fairly common and he was able to gather a number of cases of florid familial hypercholesterolemia with the classical xanthomas of skin and tendons (Figures 3.2 and 3.3).

Müller summarized his views as follows:

> The reports I have presented confirm the previous observations on xanthomatosis as a cause of hereditary heart disease. They reveal further that the syndrome of cutaneous xanthomatosis, hypercholesterolemia and angina pectoris presents itself as a well defined clinical entity. . . . There can be hardly any doubt but that xanthomatous deposits in the coronary artery and consecutive myocardial ischemia are the cause of the angina pectoris.

FIGURE 3.1　Professor Carl Müller, 1886–1983. He presented his landmark paper *"Angina pectoris in hereditary xanthomatosis"* before the Nordic Congress for Internal Medicine in 1937, and published in the *Archives of Internal Medicine* in 1939 (Source: (8)). He was the first to pull together the evidence linking familial hypercholesterolemia to coronary artery disease. Source: reprinted from reference (8) with permission.

Over the next 25 years Müller's characterization of familial hypercholesterolemia was borne out by more extensive studies of larger cohorts, especially by Wilkinson *et al.* (9), Adlersberg *et al.* (10), and Khachadurian (11). Their work established familial hypercholesterolemia as a monogenic defect, implying that the arterial disease was secondary to the elevated blood cholesterol, i.e. that the pathogenesis was analogous to the pathogenesis in Anitschkow's cholesterol-fed rabbits. Gofman and his group later showed that in patients with familial hypercholesterolemia the cholesterol elevation was all in the low density and intermediate density fractions (12). However, the nature of the gene involved remained unknown and there remained the possibility (although unlikely) that both the hypercholesterolemia and the arterial disease were determined by a single mutated gene but by two independent metabolic pathways. The elegant studies of Michael S. Brown and Joseph L. Goldstein beginning in the 1970s would dismiss that possibility by identifying the LDL receptor gene as the mutated gene and demonstrating its critical role in determining blood levels of LDL. In Chapter 5, we review in detail their Nobel Prize-winning work.

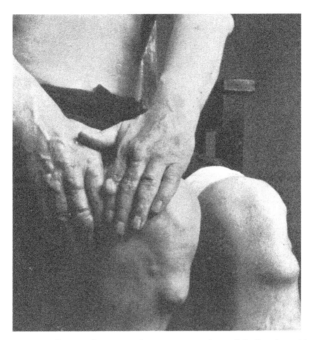

FIGURE 3.2   Severe tendon xanthomas on the extensor tendons of the hands and in the pretibial area of Müller's Case 17, a 51-year-old man with angina pectoris. His blood cholesterol level was 435 mg/dl. The patient also had xanthomas on the eyelids, elbows, and heels (see Figure 3.3). Source: reprinted from reference (8) with permission.

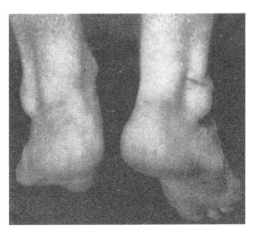

FIGURE 3.3   Enormous thickening of the Achilles tendons due to deposition of cholesterol in Müller's Case 17 (see Legend to Figure 3.2). Source: reprinted from reference (8) with permission.

DO MORE MODEST DEGREES OF HYPERCHOLESTEROLEMIA, SUCH AS ARE FOUND IN A GENERAL POPULATION, CONFER RISK OF PREMATURE CORONARY HEART DISEASE?

It could be argued – and it was argued very vigorously by many – that the concentrations of blood cholesterol in patients with familial hypercholesterolemia are so extraordinarily high – 300–400 mg/dl in heterozygotes and as much as 1,000 mg/dl in homozygotes – that it would not be legitimate to extrapolate and assume that more modest degrees of hypercholesterolemia would also increase risk. More moderate elevations might carry no risk at all. This was the same kind of argument used to trivialize Anitschkow's findings in cholesterol-fed rabbits. It is reminiscent of the arguments that surrounded the issue of whether there was a "threshold level" of radiation exposure that would carry no risk of genetic damage. In retrospect, of course, the extrapolation made sense but it would require more data on people with moderate elevations of blood cholesterol to make the case.

How High is High? What Does "Normal" Really Mean?

Today, when it is so obvious that hypercholesterolemia is critically important in atherogenesis, it is difficult to understand how so many skilled clinicians and researchers could have denied the connection between hypercholesterolemia and risk of coronary heart disease. One important reason relates to how they defined "high" and where they drew the line between normal and abnormal.

In clinical medicine the time-honored way of defining a high value for any blood component (e.g. blood glucose) is that it is any value higher than that found in 95 percent of the population (95th percentile value). So, if 95 percent of the people in the United States have a blood glucose value less than 110 mg/dl, then any glucose value above that is considered abnormal, meriting further medical work-up, but anything below that is considered "normal." This arbitrary definition of normal versus abnormal works quite well for most of the measurements that are made in the clinic. However, suppose that a particular blood component is actually known to be causing tissue damage *even at so-called normal levels* ("normal" only in the sense that 95 percent of the population have levels below it). In that case the cut-point between normal and abnormal would have to be redefined. Let's use a parable to illustrate.

AN APOCRYPHAL TALE

Before the importance of iodine in the diet was fully appreciated, there were many mountain villages in Switzerland where the diet contained insufficient iodine; consequently enlargement of the thyroid gland was very common. The enlargement was due to high blood levels of thyroid-stimulating hormone (TSH). One day an American endocrinologist visited a hospital in such a village and looked over the medical charts. He was immediately struck by how many patients had high levels of TSH and asked his Swiss colleague, "What is the normal TSH level in this town?" The clinic director said, "Between seven and eight." The American said that would be considered decidedly abnormal in the United States, where the upper limit of normal is about five. The Swiss endocrinologist drew himself up and haughtily replied, "Surely we should know what is normal for our own population! Why if we used your American values for what is normal, fully 25 percent of our citizens would be 'abnormal.' That would make no sense at all!" The American quietly told his colleague that from his walks around the town he got the impression that almost one out of every four people he encountered had a goiter. The Swiss doctors finally agreed to hold a "Consensus Conference on Lowering TSH Levels to Prevent Goiter." As a result they started a "National Iodide Education Program" and started using iodized salt. The TSH values fell and the enlarged thyroid glands disappeared!

A TRUE TALE

When the classic 95th percentile yardstick was applied to blood cholesterol levels in the U.S.A. in the 1940s, 95 percent of the population had values below 280. By the rules, then, any value below 280 was considered "normal." Most heart attacks were occurring in people with cholesterol levels well below 280, even in some patients with cholesterol levels below 200. So, if you only considered values above 280 as "abnormal" then indeed most heart attacks were occurring in individuals with "normal" blood cholesterol levels. It is understandable, then, that most clinicians considered a high blood cholesterol level irrelevant to most cases of atherosclerosis and heart attacks. One of the patients we see in our clinic had a routine check-up while he was in the Navy in 1965. He was 34 years old at the time. The laboratory reported a blood cholesterol of 380 but he was told not to be concerned and to just return to active duty! Over the next 5 years he had 4 myocardial infarctions and only began intensive cholesterol-lowering therapy around 1973. Now, at 75, he is still going strong with a negative exercise stress test. Today his cholesterol level, on treatment with a statin plus ezetimibe plus nicotinic acid, is less than 200; his LDL is about 80 and he is doing fine.

For a long time physicians simply could not accept the idea that a large fraction of the American public might have blood cholesterol levels within what was considered to be the normal range and yet be at a high risk for a heart attack. Like the Swiss endocrinologist in our parable above, American physicians balked at the suggestion that a significant fraction of our population might be "abnormal." And yet we now know that that was exactly the case: 20 percent or more of Americans with blood cholesterol levels once considered "normal" are actually working their way toward a heart attack. The results of the many recent trials with cholesterol-lowering drugs show that people in this category can sharply reduce their risk by using diet and drugs. Instead of 280, the desirable total blood cholesterol level in this country, recommended by the National Cholesterol Education Program, was originally 200 mg/dl (corresponding to an LDL of about 130). There is now good clinical trial data showing that lowering the levels still further – so that the LDL cholesterol is 70 mg/dl – further reduces risk.

THE EPIDEMIOLOGIC EVIDENCE

THE SEVEN COUNTRIES STUDY OF ANCEL KEYS' GROUP

By 1955, Ancel Keys, a pioneer in nutritional research at the University of Minnesota, was already convinced that blood cholesterol level was importantly determined by the amount and the nature of the fat in the diet (13). If blood cholesterol was a major determinant of coronary heart disease, then populations with higher blood cholesterol levels should have higher heart attack rates than other populations. He decided to launch an ambitious study in "geographic epidemiology" (14–16). Henry Blackburn, Keys' right-hand man throughout these studies, has written a fascinating "inside story" about the origins of the Seven Countries Study and the not inconsiderable logistic problems that had to be overcome (17).

Keys and his colleagues selected for study seven countries that spanned the full range of blood cholesterol levels – from Japan with the lowest to Finland with the highest. In each country (actually in several different communities in each country), blood samples were drawn for cholesterol measurement and the nature of the diet was determined by questionnaire (and, in a subset of the population, by chemical analysis); the coronary heart disease death rate was then correlated with these two variables. The average blood cholesterol in east Finland was over 260 while that in Japan was only a little over 160 mg/dl; the number of fatal heart attacks per 1,000 men over a 10-year period was about 70 in Finland and a little less than 5 in Japan. When coronary death rate was plotted against

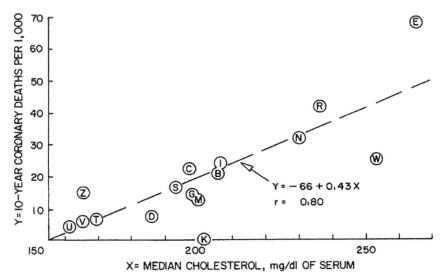

FIGURE 3.4   Coronary death rate as a function of median serum cholesterol level in Keys' groundbreaking Seven Countries Study (16). Key to symbols: B, Belgrade, Yugoslavia; C, Crevalcore, Italy; D, Dalmatia, Yugoslavia; E, East Finland; G, Corfu, Greece; I, Italian railroad workers; K, Crete, Greece; M, Montegiorgio, Italy; N, Zutphen, Netherlands; R, American railroad workers; S, Slavonia, Yugoslavia; T, Tanushimaru, Japan; U, Ushibuka, Japan; V, Velika Krsna, Yugoslavia; W, West Finland; Z, Zrenjanin, Yugoslavia. Source: (16).

the blood cholesterol level for all seven countries, the data points fell roughly on the same straight line, strongly suggesting that the population risk was roughly proportional to the blood cholesterol level over the range of values studied, as shown in Figure 3.4.

Keys' data showed also that the risk was proportional to the saturated fat intake (Figure 3.5).

The contribution of saturated fats to the total daily calorie intake in Finnish men was over 20 percent while that in the Japanese was about one-tenth of that – only about 2.5 percent. Again, the values for the other countries fell roughly along a single line. Taken together, the data showed that the risk of fatal myocardial infarction in these populations was proportional to the blood cholesterol level and proportional to the dietary intake of saturated fat.

These findings had a major impact on the cholesterol controversy. What was shown was still only correlational, but the correlation was strong and it supported the lipid hypothesis. However, it did not necessarily establish dietary saturated fat or high blood cholesterol as causal. Conceivably genetic differences or other differences in living habits might be the true explanation of the correlation. Keys and co-workers were well aware of this limitation, a limitation inherent in all epidemiological studies. Keys measured as many of the other

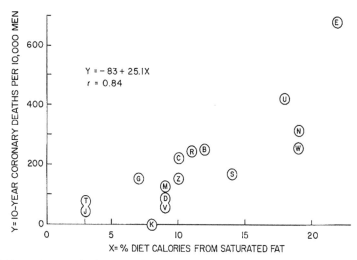

FIGURE 3.5 Coronary death rate as a function of dietary saturated fat intake (percentage of daily calories from saturated fat) in Keys' Seven Countries Study (16). Key to symbols: B. Belgrade, Yugoslavia; C, Crevalcore, Italy; D, Dalmatia, Yugoslavia; E, East Finland; G, Corfu, Greece; J, Ushibuka, Japan; K, Crete, Greece; M, Montegiorgio, Italy; N, Zutphen, Netherlands; R, Italian railroad workers; S, Slavonia, Yugoslavia; T, Tanushimaru, Japan; U, American railroad workers; V, Velika Krsna, Yugoslavia; W, West Finland; Z, Zrenjanin, Yugoslavia. Source: reprinted from reference (16) with permission.

possibly relevant factors as he could, including blood pressure, other dietary components, obesity, and many others. Even after taking these into account by appropriate statistical methods, the correlation with saturated fat intake was still significant.

THE JAPANESE MIGRATION STUDIES

How does one decide whether the differences in cholesterol levels and heart attack rates in different populations, like those studied by Keys, are really due to the differences in diet (or other environmental factors) and not due to differences in genetic make-up? A group of investigators in Hawaii hit on a clever approach to this question. Hawaii has a very large population of Japanese immigrants and so does San Francisco. The investigators determined the blood cholesterol levels and the heart attack rates in their Hawaiian Japanese population and in the San Francisco Japanese population and compared them with the same measurements in native Japanese on the island of Honshu (18). The results were striking. The Japanese who had moved to Hawaii had higher blood cholesterol levels and higher heart attack rates than the native Japanese in Honshu.

The difference was even more striking in those who had settled in San Francisco. Since the migrants studied had only been in their new environments for a few generations, there was no way that their genetic makeup could have changed significantly. The rise in blood cholesterol levels and the accompanying increase in heart attack rates following migration must have been due to environmental factors, most likely changes in dietary habits. Certainly saturated fat intake was higher in Hawaii and San Francisco than in Honshu.

THE FRAMINGHAM HEART STUDY

The Framingham Heart Study was carried out by the National Heart Institute in the small town of Framingham, Massachusetts, beginning in 1950 and continuing actively to this day (19;20). The 28,000 residents of Framingham welcomed this community-based study with open arms. A large majority of those eligible agreed to participate. Measurements were made at the baseline examination of most of the potentially relevant factors known at the time. These included blood cholesterol, blood pressure, smoking habits, obesity, diabetes, family history, and others. In later years additional measurements were added as more was learned about coronary heart disease. The initial cohort was followed with periodic exams for more than 24 years. The Framingham project was initiated by Joseph Mountain, who turned it over to Thomas R. Dawber, and it was then continued under the guidance of William Kannel and William Castelli. It continues to this day, studying the offspring of the original cohort.

Without question the data gathered in the Framingham project had more of an impact on coronary heart disease research than that from any other single epidemiological study. It provided the first solid and unarguable evidence that individuals with higher blood cholesterol levels at the time of the baseline examination were more likely to experience a myocardial infarction in the subsequent years of follow-up. It also showed that the risk was increased by a number of other factors such as high blood pressure and smoking. Moreover, the data showed that these "risk factors" were at least additive. Later studies identified additional risk factors, including diabetes, obesity, low HDL, lack of exercise, family history of coronary heart disease, and others.

These observational studies established a clear *association* between hypercholesterolemia and coronary heart disease risk but still did not establish causality. What was needed was an intervention trial, a controlled experiment that would show that lowering cholesterol levels as a single variable could reduce coronary risk. However, safe and effective drug treatment for hypercholesterolemia was still some way down the road and the effectiveness of manipulating dietary fat was only becoming clear in the late 1950s. Nevertheless, a few investigators decided to go ahead to test the lipid hypothesis on a small scale, manipulating dietary fat composition to lower blood cholesterol. What evidence did they have that dietary manipulations would indeed lower blood cholesterol levels?

DIETARY FAT, BLOOD CHOLESTEROL, AND CORONARY HEART DISEASE

The lipid hypothesis proposes specifically that a high level of cholesterol *in the blood* is a significant causative factor in atherogenesis and its clinical expression. It does not specify the nature of the hypercholesterolemia nor how it was brought about. It might be the result of genetic defects in lipid or lipoprotein metabolism; it might be the result of an endocrine disorder; or it might be the result of a diet that raises blood cholesterol levels. In the 1950s and 1960s, before effective cholesterol-lowering drugs became available, the only way to lower blood cholesterol was to manipulate the diet. Hence, there emerged a tendency to short-circuit the "diet–blood cholesterol–heart disease" problem by omitting the "blood cholesterol" and talking about the "diet–heart" problem. As a result, many observational studies examined the correlation between dietary patterns and the incidence of coronary heart disease without asking whether the dietary patterns considered actually caused a significant change in blood cholesterol levels. This resulted in a great deal of confusion and misunderstanding.

For example, the Multiple Risk Factor Intervention Trial used a modified diet designed to lower cholesterol levels, along with advice to stop smoking and advice on exercise (21). The study failed to show a significant decrease in coronary heart disease and it is often cited as a negative study that challenges the validity of the lipid hypothesis. However, the difference in cholesterol level between the controls and those on the lipid-lowering diet was only about 2 percent. This was clearly not a meaningful test of the lipid hypothesis. Even with the large numbers of men in the study, such a small difference in cholesterol level could not be expected to yield a significant result.

Another example of a study often cited as disproving the lipid hypothesis was the Minnesota Coronary Survey (22), which began in 1968 but which was not reported until 1989. Frantz and co-workers chose to do their study in six Minnesota state mental hospitals and one nursing home. A total of 4,393 men and 4,664 women were studied. They were assigned randomly to either the regular institutional diet or to a test diet with reduced saturated fat content. The achieved level of saturated fat intake in the control diet was 18.3 percent of total calories; in the experimental group it was 9.2 percent. This was enough of a change to reduce the mean serum cholesterol level by a little over

15 percent.[1] However, there was no difference in cardiovascular events or cardiovascular deaths between the two groups. What is often lost sight of, although duly pointed out by the authors, is that there was an enormous rate of turnover of the patients in the mental hospitals studied. They noted that by the time the study got under way "the practice of vigorous drug treatment [of mental illness] and early discharge to the community was in full swing." As stated in the Abstract: "The mean duration of time on the diets was 384 days." Only 17 percent of the subjects had been on the test diet for more than 2 years! In the Lipid Research Clinics study (23;24) and in the Helsinki Heart Study (25) favorable trends did not appear for at least 2 years. Clearly, the Minnesota Coronary Survey cannot be regarded as a meaningful test of the lipid hypothesis. Yet it is often included in meta-analyses and often appears on lists of studies allegedly disproving the lipid hypothesis.

The relationship between diet and blood cholesterol is complex; many different dietary components can affect blood cholesterol levels (protein, carbohydrate, total calories, plant sterols, etc.). By far the most controversial component has been the fat content – both the quantity of fat and its composition. The American Heart Association and the American Medical Association began as early as 1961 to recommend decreasing daily intake of saturated fat and replacing it with polyunsaturated fat (26). They also recommended reducing total fat intake and cholesterol intake. Those recommendations were seriously questioned at the time and they have been repeatedly attacked even to this day.

What was the scientific rationale for those early recommendations? Were they consonant with the knowledge base at the time? And why were views on the wisdom of these recommendations so discordant?

DO DIETS RICH IN SATURATED FATS REALLY RAISE BLOOD CHOLESTEROL?

The epidemiologist approaches this question by looking for correlations between animal fat intake and blood cholesterol levels in large populations. The experimentalist studies individuals (or small groups of individuals), changes the proportion of animal fat in their diet (hopefully as a single variable), and looks for changes in blood cholesterol levels. The results of these two approaches may not agree. As nicely pointed out by Blackburn et al. (27;28), if there is a good deal of variability from individual to individual in the magnitude of the blood cholesterol response to changes in saturated fat intake or if a number of other variables help determine blood cholesterol levels, a correlation may not be seen in the *observational* study. Yet that is not necessarily in conflict with an *experimental* finding that *individuals*, when shifted from a diet rich in polyunsaturated fats to one rich in saturated fats, all show an increase in blood cholesterol level but to differing degrees. Failure to recognize the important difference between *observational* and *experimental* findings accounts for a lot of the confusion in this area.

EXPERIMENTAL FINDINGS

Pioneering Studies from the Netherlands

Long before the relevance of blood cholesterol to heart disease was suspected, a Dutch physician, C.D. deLangen, posted to the Dutch East Indies as a public health officer, demonstrated the role of diet for the first time. In 1916, shortly after cholesterol was first characterized chemically and could be measured easily, he reported that the blood cholesterol levels of the natives in Indonesia were considerably lower than those of the Dutch colonists. He speculated that this might be due in part to the very rich diet of the Dutch compared to the much more spartan diet of the natives. The natives subsisted mainly on vegetables and rice while the Dutch colonialists enjoyed a rich butter-eggs-and-meat diet (29). In 1922, he did what was probably the first controlled diet–heart study. He put five Indonesian natives on a cholesterol-rich diet like that of the Dutch colonists (rich in eggs and meat). After 3 months their blood cholesterol level had increased by an average of 27 percent. He also reported that Indonesians who had migrated to Amsterdam had cholesterol levels just as high as those of their Dutch counterparts. Clearly the lower cholesterol levels of the Indonesians in Indonesia were not based on genetic differences. In Amsterdam the émigrés had adopted the ways of the host country and diet was probably the major factor (30). DeLangen's work was published only in Dutch and in a rather obscure Dutch journal. It is seldom cited but he anticipated correctly the results of the more extensive studies done 30 years later. Another opportunity missed.

The Dutch, for some reason, seem to have been fascinated by the diet–blood cholesterol relationship. Just after World War II, it was another Dutchman, J. Groen, who called attention to the striking decrease in heart attack rate during the war. He pointed out that this coincided with severe shortages of some foods, in particular fatty foods. He proposed that this was not a coincidence but that the unavailability of animal fats in particular had caused a decrease in blood cholesterol and that this in turn caused the observed decrease in the risk of heart attacks. The fact that heart attack rates returned to pre-war levels a few years after the war supported his hypothesis. Quite obviously, however, many other things were also changing besides the availability of dietary fat. The hypothesis was tenable but the evidence was circumstantial at best. However, after the war Groen added strength to his hypothesis by carrying out carefully controlled human trials showing the effects of diet on blood cholesterol levels (31).

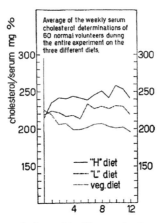

FIGURE 3.6   Changes in serum cholesterol in 60 normal volunteers studied by Groen *et al.* in 1952 during 12 weeks on one of three diets differing mainly in animal fat content, The "L" diet approximated the usual Netherlands diet of the time; the "H" diet increased the amounts of animal fat, cheese, and eggs; the "V" or "veg" diet was a strictly vegetarian diet except for 100 g of milk daily and *ad lib* skim milk or buttermilk. As shown, the serum cholesterol on the "veg" diet fell to below 200; on the "H" diet it rose to 240–250; on the "L" diet it fluctuated and perhaps rose slightly toward the end of the test period. Source: (31).

In 1952, Groen and colleagues worked with a group of 60 healthy normal volunteers. All meals were taken in a common dining room under supervision of dietitians. Each subject rotated among three different diets, following each one for 12 weeks at a time. The diets were: "V", a strictly vegetarian diet (zero meat, fish, eggs, milk, cheese, cream, or butter); "L", a typical Netherlands diet (50 g of meat, 30 g of cheese, 0.5 liters of milk daily); and "H", what they called a "standard American" diet, containing *at least* 250 g of meat, 50 g of cheeses, and unlimited amounts of milk, cream, butter, and bread. As shown in Figure 3.6, the mean serum cholesterol for those on the typical Netherlands diet (L) changed very little from the baseline value of 225 mg/dl over the 12 weeks on that diet; the value for those on the vegetarian diet (V) fell to slightly below 200; and for those on the "standard American" diet (H) the value rose to over 250 mg/dl. Note that the level reached on the vegetarian diet was fully 25 percent lower than that on the "standard American," animal fat-rich diet (H).

Groen looked carefully at the results in the different individuals in his cohort and made the important observation that there were considerable differences in "the reaction of each subject's serum cholesterol to the diet." One subject, with an initial cholesterol level of about 180 mg/dl showed no rise in level on the "H" diet and not much of a rise even when he increased his meat intake to 500 g/day. Groen said he was "an example of what one might be inclined to call a 'constitutional hypocholesterolemia.'" Today we would designate him a "nonresponder."

As discussed below this rather wide variance in response to diet probably accounts for the apparent absence of correlation between serum cholesterol levels and dietary fat in some within-country observational studies.

RIGOROUS METABOLIC WARD STUDIES

Carefully controlled metabolic ward studies by four American laboratories beginning in 1952 documented the importance of saturated fat in the diet as a contributor to hypercholesterolemia. These were the groups of Kinsell *et al.* (32), Ahrens *et al.* (33), Keys *et al.* (34), and Beveridge *et al.* (35–37). All of these investigators carried out metabolic ward studies in volunteer subjects under close surveillance. They did single-variable experiments, i.e. they kept all the elements in the diet constant except that a saturated fat was substituted for a polyunsaturated fat (or vice versa). The total fat content was not changed and there was no change in body weight. Both Kinsell (38) and Ahrens (39) utilized the elegant tool of the liquid formula diet in some of their studies, i.e. the subjects took *all* their diet in the form of a "milk shake" of precisely known composition given orally several times daily. Some of the studies lasted months and the subjects understandably got a bit bored with their liquid cuisine. In an attempt to ameliorate the monotony of this regimen, the investigators offered the volunteer subjects their choice of flavorings ... chocolate, strawberry, etc. Some patients opted for the flavor of their choice but, surprisingly, most elected to return to the unflavored formula within a short time! Flavored or not, the liquid formula diet was monotonous in the extreme but it was the perfect tool for insuring that dietary intake was constant and quantitative. The Ahrens approach was later widely adopted as the gold standard for precise metabolic ward studies.

The results were clear-cut: when the formula contained an unsaturated fat (e.g. corn oil or safflower oil) the blood cholesterol level fell from the level on an ad lib diet; when that unsaturated fat was replaced by a calorically equivalent amount of a saturated fat (e.g. butter, lard, or coconut oil), *changing nothing else in the formula*, the blood cholesterol rose. Each subject served as his or her own control so there was no confusion due to individual idiosyncrasies in response (40). An example of the results of such a study are shown in Figure 3.7.

Note that the cholesterol level on switching fats plateaued within a few weeks and then remained fairly stable until the next switch in diet composition. In a few cases, the effect of a given fat substitution was studied several times in the same subject and the results were highly reproducible. However – and this is highly relevant to the later discussion – the magnitude of the effect of a given fat substitution varied considerably from individual to individual.

In Ahrens' studies of more than 40 subjects, some of them hypercholesterolemic, the isocaloric substitution of butter fat or coconut oil for corn oil

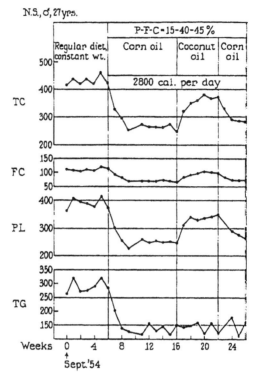

FIGURE 3.7   Responses of a hypercholesterolemic patient to changes in the nature of the fat contained in a liquid formula diet (36). After 8 weeks on an *ad lib* diet, liquid formula was started as the only source of calories with 15 percent coming from protein, 40 percent from fat and, 45 percent from carbohydrate. On starting the corn oil-rich diet, serum cholesterol fell from over 400 to below 300 mg/dl and remained there for 8 weeks. On substitution of an equicaloric amount of coconut oil, cholesterol level rose sharply, almost back to the 400 level. Returning to the corn oil formula caused the level to fall again, down to below 300. Source: (40).

raised the total cholesterol level by 25–50 percent (40). Substitution of palm oil raised it by 10–30 percent. Qualitatively similar results were obtained by Beveridge *et al.* (41), by Bronte-Stewart *et al.* (42), and by Keys *et al.* (34), using different methods but all arriving at the same basic conclusion: substitution of polyunsaturated fats for saturated fats, *other factors being held constant, including total fat intake*, reduces blood cholesterol levels. Keys stated the case nicely in 1957: "it is clear that it is unnecessary to prescribe a diet extremely low in *total* fats to lower the serum cholesterol; exclusion of the *saturated* fats (in butterfat and meat fats) has the greatest effect, and this effect may be enhanced by substitution of such oils as corn oil and cottonseed oil" (43).

Another source of confusion in evaluation of dietary manipulations has been the failure to take into account the nature of the reference diet. What were the patients switching *from* and what were they switching *to*? If the subjects studied are on a high saturated fat diet to start with, as most Americans are, they will show a nice drop when switched to a polyunsaturated fat-rich experimental diet. On the other hand, if they are already on a polyunsaturated fat-rich diet there may be little change. Also, there are many additional factors, including other dietary factors, which can affect blood cholesterol levels (e.g. caloric balance, plant sterol intake, the nature of the proteins). Changing diets to alter saturated fat intake may involve, intentionally or unintentionally, significant changes in some of these other factors as well and therefore modify the response. That is why cross-sectional studies within a fairly homogeneous population, e.g. the Framingham population, have sometimes failed to show a correlation between dietary fat patterns and serum cholesterol, as nicely pointed out by Jacobs *et al.* (27) and by Blackburn and Jacobs (28).

UPDATE ON THE EVIDENCE FROM METABOLIC WARD STUDIES

Before going on, I would call attention to a recent meta-analysis of 395 published metabolic ward studies that have tested the extent to which replacement of saturated fats by complex carbohydrates reduces total blood cholesterol and LDL cholesterol. Pooling the data from these studies, Clarke *et al.* (44) found that total cholesterol fell on average by about 21 mg/dl and LDL cholesterol fell by about 14 mg/dl. The pooled results also showed that reducing dietary cholesterol by 200 mg per day would on average reduce blood cholesterol by about 5 mg/dl. They calculated that if the British could replace 60 percent of the saturated fat in their diet by complex carbohydrates or other fats and avoid 60 percent of the cholesterol in their diet, they would reduce their blood cholesterol by 32 mg/dl – by about 10–15 percent. At first glance this may not seem like very much but a 10 percent drop in blood cholesterol translates into a 20–30 percent reduction in cardiovascular events. In the U.S.A., with almost 500,000 deaths per year from coronary heart disease, that would correspond to 100,000–150,000 lives saved.

The pioneering nutritional studies outlined above left no doubt that the nature of the fat in the diet, saturated versus polyunsaturated, and the amount of cholesterol in the diet, at least under some circumstances, could significantly affect blood cholesterol levels in individual healthy volunteers. These studies were done under carefully controlled conditions, most of them in research hospital metabolic wards, and on a limited number of subjects. Could the results be extrapolated to "the real world?" Would the differences observed in the research setting in a handful of subjects be seen also in the general population? And, most importantly, if a significant drop in blood cholesterol level was achieved and maintained would there be a decrease in heart attack rates?

These metabolic ward studies did not directly address the question of whether the *total* fat content, independent of composition, might be a determinant of blood cholesterol levels. Keys' original data had shown a positive correlation of both total fat intake and saturated fat intake with coronary heart disease risk but he later concluded that the correlation with total fat intake could be largely due to the parallel accompanying increase in the intake of *saturated fat*. There continues to be controversy about the importance of total fat intake, largely due to misunderstandings of the points made above, but there is concurrence on the value of reducing intake of saturated fats.

In any case, these metabolic ward studies provided the scientific rationale for the dietary intervention studies that followed soon thereafter, asking if lowering plasma cholesterol by increasing polyunsaturated fat intake would reduce coronary heart disease events (45–47). The effects on coronary heart disease risk in these dietary intervention studies are discussed in a later section of this chapter. Here we simply want to note that the results of these intervention studies confirmed, in the more than 3,000 men studied, that replacing saturated fat by polyunsaturated fat does indeed decrease blood cholesterol level – by 17.6 percent in the Leren study (45), by 12.7 percent in the Dayton study (46), and by 12–18 percent in the Miettinen study (47). These highly significant effects on blood cholesterol are often lost sight of, probably because the primary focus in the studies was on whether there would be a significant change in coronary heart disease risk. The confirmation in a large number of subjects from a general population that the *nature* of dietary fat, independent of the *amount* of fat, affects plasma cholesterol level was an invaluable by-product of these pioneering intervention trials.

THE NATIONAL DIET–HEART STUDY: DIRECT EVIDENCE ON A LARGE SCALE IN A GENERAL POPULATION THAT SUBSTITUTING POLYUNSATURATED FAT FOR SATURATED FAT REDUCES BLOOD CHOLESTEROL LEVELS

As early as 1960, the National Heart Institute began to explore the possibility of critically testing the lipid hypothesis in a large-scale dietary intervention study. An Executive Committee under the chairmanship of Irvine H. Page worked for 2 years to develop a proposal that was approved and funded in 1962. The six principal investigators were Benjamin M. Baker, Johns Hopkins

University; Ivan D. Frantz, Jr., University of Minnesota; Ancel Keys, University of Minnesota; Laurence W. Kinsell, Highland-Alameda Hospital; Jeremiah Stamler, Chicago Board of Health; and Fredrick J. Stare, Harvard University School of Public Health. The long-term goal was to have been a huge national study in a general population to determine whether reducing blood cholesterol by diet in a double-blind design would reduce coronary heart disease events. That full-scale study never got off the drawing board but the first phase, feasibility study, was successfully completed. The results were published in 1968 and provided new and compelling evidence that substitution of polyunsaturated fat for saturated fat lowers blood cholesterol in a general population (48). It was a truly heroic undertaking in which food manufacturers provided specially processed foods in which the saturated fat content was manipulated. Participants ordered their fat-containing foods from a central depot and were given dietary instructions. The success of the double-blind design was established by a post-study questionnaire.

Men aged 45–54 were recruited by census tract and were randomly assigned to one of three diet groups. This was a general population, not a high-risk, hypercholesterolemic group. Mean total serum cholesterol was 230 mg/dl. Three basic diets were studied: control "D", a diet that mirrored the average American diet with 40 percent of total calories from fat and a polyunsaturated to saturated fat ratio (P/S) of 0.4; experimental "C", high P/S ratio (2.0) but same total fat content (40 percent of calories); and experimental "B", lower total fat content (30 percent of calories) and a P/S ratio of 1.5. Almost 400 men were studied for a year on each of these three diets. The nominal control group, "D", instead of showing the anticipated neutral response showed a 2.4 percent drop in serum cholesterol. The men in group "C" showed an 11.7 percent drop, and those in group "B" a 10.8 percent drop. Adherence was undoubtedly less than complete as shown by the results in a parallel closed study in a mental hospital population using exactly the same three diets. There the drops in cholesterol levels were significantly larger: 15.0 percent on diet "C" and 16.5 percent on diet "B". Table 3.1 shows the wide spread in individual results on the same diet. On diet "B", 15.0 percent of the men dropped their serum cholesterol by 29 percent or more while 27.2 percent only dropped their levels by less than 5 percent.

Based on the degree of cholesterol lowering in the open populations the investigators estimated that 50,000–100,000 healthy men in the age range 40–59 would need to be recruited and followed for 5 years to test adequately the lipid hypothesis. Unfortunately, that study was never done. The cost would have been enormous, forcing a slow down in all of the other research programs of the NIH. Moreover, there was no guarantee that the result would be clear-cut. In retrospect, it might have been wiser to give up on the double-blind design, something the Executive Committee did consider at one point, and/or to limit the cohort to men at high risk, thus markedly reducing the cost of the study.

TABLE 3.1   Individual responses to a diet low in saturated fat and enriched in
polyunsaturated fat*

Diet group	Percentage of subjects with the indicated percentage drop in serum cholesterol**				
	>20%	15–19%	10–14%	5–9%	<5%
B	15.0	17.7	19.3	20,8	27.2
C	16.6	20.3	20.6	18.3	24.1

*Data from the National Diet–Heart Study (48). Diet in Group B provided 30% of total calories as
fat with a polyunsaturated/saturated fat ratio of 1.5; diet in Group C provided 40% of total calories
as fat with a P/S ratio of 2.0; diet in the reference group, Group D, provided 40% calories from fat
with a P/S ratio of 0.4, approximating the fat in the usual American diet.
**The percentage drop uses the serum cholesterol values in the men in group D as the referent.
There were over 300 men in each group and the diets were followed for a 1-year period.

The key point is that the Diet–Heart Study provides data in about 1,000 free-living men drawn from a general population showing that blood cholesterol level is significantly higher when the saturated fat intake is higher. The contention that there is no hard science supporting this view (49) ignores this national study and all the other experimental demonstrations of the relationship summarized above.

OBSERVATIONAL STUDIES

In contrast to the persuasive experimental findings just reviewed, within-population observational studies have frequently failed to find a significant correlation between total fat intake or saturated fat intake and plasma cholesterol levels. That was true, for example, in the Framingham Study (50) and in the Tecumseh Study (51). Why the discrepancy?

One explanation is that the variability from person to person in terms of response to dietary factors is so great that it dilutes out the correlation. If the subjects making up a cohort were studied individually, as in one of Ahrens' metabolic ward protocols, almost all would show the same qualitative responses but to different degrees. Jacobs et al. dealt with this issue formally in a 1979 paper (27) and a 1984 editorial (28). They pointed out the large variability encountered from individual to individual in responses to changes in dietary fat. The data in Table 3.1 illustrate the point. Jacobs et al. used a mathematical model to show that a zero correlation is what you would actually expect. They also applied their model to a set of experimental data and showed that, again, a zero correlation might be expected. In short, a zero correlation in a population study does not necessarily negate the possible role of dietary fat in helping to determine plasma cholesterol levels in individuals. Moreover, plasma cholesterol levels are affected by many variables in addition to dietary fat intake, which contributes further to the dilution.

THE CONTRIBUTION OF DIETARY CHOLESTEROL INTAKE

The cholesterol content of the diet also makes a difference. However, the effect of increasing cholesterol intake is generally much less impressive than the effects of increasing saturated fat intake. The reason for this, as first shown by Dr. William Connor and his colleagues, is that beyond a certain "threshold" level of cholesterol intake further increases have little additional effect on blood cholesterol level (reviewed in (52)). Adding just 100–200 mg cholesterol per day to a previously *cholesterol-free* diet can raise the blood cholesterol level by as much as 20 percent. Once the total daily cholesterol intake has reached about 300 mg per day there is a "saturation phenomenon" such that further increases in cholesterol intake have very little effect on blood cholesterol levels. In a sense the damage has already been done when the cholesterol intake is at or above 300 mg. For most Americans, daily cholesterol intake is already at or above 500 mg/day on their usual diet. Therefore, adding more cholesterol, even very large amounts of it, may not cause much of an increase in blood cholesterol level. That is probably why some researchers have found no effect of adding a couple of eggs per day to a basal diet (one already containing more than 300 mg/day), while others have found adding eggs to have a significant cholesterol-raising effect (when the reference diet is low in cholesterol). Similarly, reducing an already very high intake, say 800 mg/day, by 100 or 200 mg/day, may not lower blood cholesterol significantly. To lower blood cholesterol levels significantly in such subjects the cholesterol content of the diet may need to be lowered to 300 mg per day or less. Moreover, the responses of individual subjects are not easy to predict. Different people absorb dietary cholesterol at very different rates and catabolize it at very different rates. One extreme example appeared as a case report describing an 88-year-old man who through much of his life consumed 25 eggs every day yet maintained a blood cholesterol level between 150 and 200 mg/dl (53). Careful metabolic research studies showed that his body was converting the dietary cholesterol into bile acids with great efficiency. Rats and some other species use the same "trick" to keep their blood cholesterol levels low even when their dietary intake of cholesterol is high.

While the magnitude of the changes induced by changes in cholesterol intake in humans is much less than that accompanying changes in saturated fat intake, a recent meta-analysis shows that dietary cholesterol significantly raises the ratio of total cholesterol to HDL cholesterol and is thus proatherogenic (54).

EARLY CLINICAL TRIALS OF DIETARY INTERVENTION

THE PAUL LEREN OSLO STUDY, 1966

Almost as soon as it was reported that diets rich in polyunsaturated fats and low in cholesterol could lower blood cholesterol levels, a young physician in Oslo, Paul Leren, started planning the "next step" study. In 1957, he ran a pilot study to see how much of a decrease in blood cholesterol level could be obtained by dietary means and whether it could be sustained. The key element of the diet was a sharp reduction in saturated fat and cholesterol intake and an increase in polyunsaturated fat intake. In fact, each experimental subject had to consume a pint of soybean oil every week, adding it to salad dressing or using it in cooking or, if necessary, just gulping it down! Leren bravely launched his 5-year study with 412 myocardial infarction survivors, counting on their high level of motivation and intensive reinforcement from dietitians to keep them compliant and they were. Sixty percent of the men were considered to be "Excellent" adherers and their blood cholesterol level fell from an average starting value of 296 mg/dl to an average of 232 mg/dl during the course of the study – a drop of 21.6 percent. Adherence by the rest of the men was lower so the mean drop in cholesterol for the group as a whole was 17.6 percent (Figure 3.8).

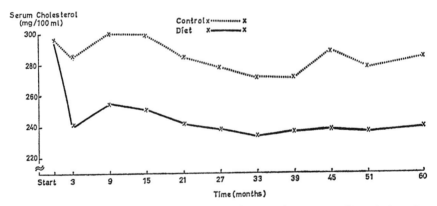

FIGURE 3.8   Serum cholesterol levels in control and diet treated groups in Paul Leren's pioneering 1966 study showing that cholesterol lowering by substitution of polyunsaturated fat for saturated fat reduces risk of myocardial infarction in men who had had a previous infarction (42). Note that the initial levels were close to 300 mg/dl in these men with advanced coronary heart disease. It fell promptly – by a little over 20 percent – on starting the experimental diet and remained well below that of the controls for the 5-year duration of the study. It was this unusually large drop in cholesterol level that enabled Leren to get a statistically significant 37 percent protective effect even with only 206 men in each group. Source: reprinted from reference (44) with permission.

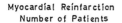

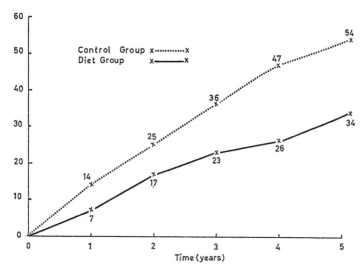

FIGURE 3.9   Myocardial reinfarctions over the 5 years of Paul Leren's 1966 study of dietary cholesterol lowering to reduce coronary heart disease risk (cf. Figure 3.8). Source: reprinted from reference (44) with permission.

The key finding was that 54 patients of the 206 in the control group (26 percent) had a second heart attack during the 5 years of the study compared with only 34 of the 206 in the diet group (16 percent), see Figure 3.9.

The result was significant at the $p < 0.03$ level (45). A follow-up of the 412 man cohort at 11 years (55) showed a strikingly lower myocardial infarction mortality in the treated group ($p < 0.004$).

One criticism of Leren's study was that there was no difference in all-cause mortality but the number of subjects that would have been needed in order to detect a statistically significant decrease in all-cause mortality would have been much larger.

Here was a carefully conducted study reported in 1966 with a statistically significant reduction in reinfarction rate. Why did it not receive the attention it deserved?

THE WADSWORTH VETERANS ADMINISTRATION (VA) HOSPITAL STUDY, 1969

The Wadsworth VA Hospital in Los Angeles operated a domiciliary facility where healthy but needy veterans could reside at no cost. They took almost all their meals in one of two dining halls on the premises. In the late 1950s, Seymour Dayton, Morton L. Pearce and their co-workers saw this as an ideal setting for a test of the effects of unsaturated fat on atherosclerosis (46). All the men in the study were assigned to either dining hall A or dining hall B. Dining hall A would continue to serve the usual diet but dining hall B would serve a modified diet. The main difference was that in dining room B vegetable oil, rich in polyunsaturated fat, would be substituted for about two-thirds of the animal fat. The total fat content of the two diets, however, would be kept the same, providing 40 percent of total calories. About 800 men, most of them in their 60s or 70s, were randomly assigned to one or the other dining room and followed for up to 8 years. To test whether the physicians examining the men and evaluating clinical outcomes knew to which group they had been assigned, the physicians were asked to fill out a questionnaire near the end of the study. The percentage of correct assignments was 49–54 percent, just what you would expect by chance alone.

An objective measure of adherence came from analysis of the fatty acids in adipose tissue biopsies. Samples were analyzed at the beginning of the study and again after 5 years. The shift to polyunsaturated fatty acids in the samples from most men in the experimental group was very close to that predicted if there had been good adherence to the diet. Most of the men, but not all, were free of heart disease at the beginning of the study.

The blood cholesterol level in the men eating in dining room B fell promptly after the switch in diets and continued to be lower than that for the men eating in dining room A – mean difference 29.5 mg/dl, or 12.7 percent. Neither group showed any significant change in body weight.

The number of combined events (definite heart attack, fatal or nonfatal; stroke; or peripheral atherosclerosis requiring amputation) was reduced by 31 percent in the experimental group (48 versus 70) and that difference was statistically significant by the usual convention ($p < 0.05$). However, at the beginning of the study the definition of "hard" end points had not included stroke or advanced peripheral arterial disease. If those were excluded the difference in event rate was reduced to 18 percent and that difference did not reach significance.

This study, along with the Finnish Mental Hospitals Study discussed below, stands as one of the most important and persuasive studies of the pre-statin era. It was a randomized trial of a very clever design and it is difficult to fault the planning and execution. Yet, if you evaluate it rigorously in the standard way – by asking if the end points as initially defined in the protocol showed a statistically significant effect – the answer is "No." If you add stroke to the initially defined end point then there is a significant effect ($p < 0.02$). At the time it was thought that strokes did not necessarily follow the same pattern as myocardial infarction. Today we know that aggressive lowering of cholesterol levels (with statins) does reduce stroke incidence as well. However, under the rules of the statistical game in 1969 that was not allowed. Result? This superbly conducted study got short shrift.

THE FINNISH MENTAL HOSPITALS STUDY, 1968

At almost the same time that the Wadsworth VA study was getting underway in Los Angeles, a group in Finland was planning a study using a very similar approach (47;56). However, instead of two separate dining halls in a single institution they used two separate psychiatric hospitals, leaving the diet at one hospital (hospital N) unchanged, but introducing a polyunsaturated fat-rich diet at the other (hospital K). The major diet changes were the use of "filled milk" (replacement of milk fat by soybean oil) and substitution of a special polyunsaturated fat-rich margarine for butter. The usual Finnish diet at the time was extraordinarily rich in saturated fat and this was reflected in the blood cholesterol levels, averaging about 270 mg/dl at both hospitals at the beginning of the study.

The meaningfulness of this study was greatly enhanced by the use of a "cross-over" design. After 6 years, the diets at hospitals N and K were switched. For the next 6 years the patients in hospital N now ate the polyunsaturated fat-rich diet and the patients at hospital K (probably with great sighs of relief) went back to their familiar fatty foods. This was a very large study with almost 30,000 person-years of follow up. Furthermore, the study was blinded in the sense that the review of causes of death was carried out by physicians who did not know from which hospital the patients came.

On the experimental diet, blood cholesterol levels were 12–18 percent lower than they were on the standard Finnish fare. For example, the level in the men in hospital K, which continued on standard Finnish fare during the first 6 years, averaged 268 mg/dl; at hospital N it was 217 mg/dl. As in the Wadsworth VA study, fat tissue biopsies were analyzed and these verified adherence to the diet. Over 2,000 patients were involved but not all were in the hospital for the full 12 years of the study.

There was a strikingly lower death rate from coronary heart disease on the experimental diet. Among the men it was reduced by about 50 percent ($p < 0.001$). The results in women were in the same direction but reached statistical significance only among the women in hospital N ($p < 0.001$).

OVERVIEW OF THE THREE LARGE PRE-1970S TRIALS

These three studies followed over 3,600 subjects for 5–12 years. All three involved the reduction of serum cholesterol by substituting polyunsaturated vegetable fats for saturated animal fats, i.e. they reduced the relative and absolute daily intake of saturated fatty acids and cholesterol and simultaneously increased the daily intake of polyunsaturated fats. In all three studies, the drop in serum cholesterol was substantial with these aggressively modified diets: −21.6 percent in Leren's Oslo study; −12.7 percent in the Wadsworth VA study; and −12–18 percent in the Finnish mental hospitals study. A statistically significant decrease in coronary events was seen in two of them; the decrease in the VA study was appreciable but reached significance only when pooled with other cardiovascular end points.

ADDITIONAL PRE-1970 STUDIES

THE LESTER MORRISON STUDY, 1955

Lester J. Morrison was a private practitioner of cardiology in Los Angeles. He was one of the few who took quite seriously the implications of the animal experiments of Anitschkow. He decided as early as 1946 that lowering blood cholesterol might be therapeutic and began what was probably the first study testing the possible benefit of cholesterol lowering (57;58).

The design was very simple: every other patient referred to him after a heart attack was assigned to a low-fat, low-cholesterol diet while the alternate referrals were told just to continue their customary diet. There were only 50 patients in each group, mostly men, and the mean age was 61. The experimental diet was spartan – only about 25 g of total fat and only 50–70 mg of cholesterol daily – more rigorous even than the diet currently recommended by the American Heart Association. But these men were very highly motivated, having just recovered from a heart attack. The blood cholesterol level in Morrison's experimental group fell from 312 mg/dl to 220 mg/dl – almost a 30 percent change – reflecting their motivation! After 8 years of observation 38 of the 50 patients in the control group had died but only 22 in the diet-treated group had died, a dramatic result indeed.

A major problem with this study is that, like most dietary trials, it was of necessity not double blinded. Both Dr. Morrison and the patients knew to which group they were assigned, making it not unlikely that the caregivers might (albeit subconsciously) lavish more TLC on those in the experimental group or pay more attention to their blood pressure and so on. In that connection it is relevant that the patients on the experimental diet lost an average of 8–10 kg. The study group was small and the report did not adequately compare the groups with regard to other risk factors, nor were the criteria for defining events described in sufficient detail.

Some investigators felt this was a "too-good-to-be-true" study. When the results of later dietary trials began to come in, reporting much more modest decreases in blood cholesterol level, that feeling was reinforced. In any case, Morrison's results were dismissed by most people in the field as a "fluke" (or worse).

In retrospect, Morrison's patients may have been more like those treated recently by Nathan Pritikin (59) or by Dean Ornish (60), using an almost fat-free diet and prescribing intensive exercise and weight loss. On that regimen, patients do show remarkable drops in cholesterol levels and reductions in blood pressure, and some are reported to show actual regression of lesions, documented by coronary angiography.

THE ANTI-CORONARY CLUB STUDY, 1966

In 1957, the same year Leren started his study in Oslo, the Bureau of Nutrition of the New York City Department of Health began a very similar study (61). They studied a group of 814 men free of coronary heart disease (CHD) at the beginning of the study but at high risk because of elevated blood cholesterol. Most of the subjects also had at least one additional risk factor – high blood pressure or obesity. As in Leren's study, the diet was low in total calories from fat and very low in saturated fat and cholesterol. A little over 800 men completed the study, which spanned 7 years. The cholesterol level in the experimental group fell by 13 percent while that in the control group remained unchanged. During the first 2 years there was no significant difference between the two groups in the rate of appearance of new CHD but after an additional 2–3 years of follow-up the difference was large (more than a 60 percent reduction in event rate) and statistically significant (p-value, 0.01). Certainly a dramatic result – but the study was seriously flawed.

First, the control group was recruited from a quite different population – men who had volunteered for examination at a cancer clinic. So baseline characteristics may have been different. Second, the "new cardiovascular events" included soft end points – angina pectoris and electrocardiographic changes not necessarily diagnostic of CHD. Third, the total number of subjects studied was small and the number of events was unexpectedly small – 8 out of 814 in the experimental group and 12 out of 463 in the controls. Still, the difference was statistically significant at the $p < 0.01$ level.

Here we have a study with a number of important weaknesses. Even though the end result was statistically significant, if this were the only study available few researchers would have been persuaded. Again, however, the result was positive and statistically significant, deserving some consideration in assessing the lipid hypothesis.

THE BIERENBAUM ST. VINCENT'S HOSPITAL STUDY, 1967

Bierenbaum *et al.* (62) studied 100 young men (20–50) with prior myocardial infarction on fat-modified diets for 5 years They actually used two different diets but the cholesterol-lowering was equal on the two and so the data were pooled. Total serum cholesterol decreased by 9 percent. The controls were a cohort of myocardial infarction survivors chosen to match the experimental group in relevant baseline characteristics, i.e. this was a case-control study, not a randomized trial. Over the 5-year study period the recurrence rate for infarctions in the experimental group was 4.4 percent and in the controls 7.1 percent, a statistically significant 38 percent reduction ($p < 0.01$).

Again, a limited study with serious flaws but statistically significant and supportive of the lipid hypothesis.

THE BRITISH MEDICAL RESEARCH COUNCIL (MRC) STUDY, 1968

This study, carried out in four cooperating London hospitals, involved about 400 men who had had their first heart attack very recently. They were randomly assigned to a control group or to an experimental group. The latter were instructed to cut back on saturated fats and to consume 3 oz of soya bean oil daily. The end point was the first major relapse, defined as a definite second heart attack, fatal or nonfatal. Other nonfatal relapses were rapidly worsening angina or heart failure due to a new heart attack. Half the men were in the trial for 4 years or more, the other half for fewer (63).

Serum cholesterol fell by 33 percent initially, but by the fifth year it was only 12 percent below the baseline value. There was no statistically significant difference in event rates in the two groups although there were more events in the control group (74 versus 62; nonsignificant).

The major problem with this study is the small sample size. In their Discussion the authors point out that with the limited number of patients enrolled (400) they would have detected a significant result *only if there had been a 50 percent decrease in event rates in the experimental group*. To be sure of detecting with confidence even a 25 percent reduction of all relapses they would have had to enroll four times as many patients, i.e. 1,600 instead of 400. So the "negative" result here does NOT prove that the diet is without effect; it only says that the effect, if any, is less than a 50 percent reduction in event rates. This is a major problem with most of the early studies. The MRC investigators point out that their study design and the characteristics of their patients were very similar to those in the Leren study in

Oslo (64). In fact when they combined the results of the two studies there was a 31 percent reduction in nonfatal myocardial infarctions.

OVERVIEW OF ALL OF THE PRE-1970S DIET INTERVENTION STUDIES

Looking over the studies described above, what can one reasonably conclude? Yes, some of the studies, particularly the Leren study, the Wadsworth VA study, and the Finnish Mental Hospital study, were positive and statistically significant. Others, while showing a trend toward protection, did not reach statistical significance. The British Medical Research Council Study showed no benefit at all. However, as discussed above, the numbers of subjects in that study and the degree of cholesterol lowering were such that they would have demonstrated statistically significant benefit only if the diet had reduced risk by fully 50 percent! Even the positive studies involved rather small numbers of subjects, and not all of them satisfied the statisticians' conventional criteria for significance. Taking an overview of all the trials, one would say the data were certainly suggestive but by no means airtight. Was the evidence strong enough that physicians should have started recommending dietary changes to their patients? Strong enough to justify a national program to get people to change their diets?

Most practicing physicians were less than impressed. Partly this reflected their weighing of the evidence but partly it may have reflected their pessimistic feeling that getting people to change their diets radically was a vain hope.

Others felt that the data were indeed sufficient. They based their conclusions not only on the formal clinical trial data but also on the wealth of data from studies in animal models of atherosclerosis, epidemiological studies, and extrapolation from the experience with patients having extraordinarily high cholesterol levels due to inherited abnormalities. They contended that the overall information available more than justified a recommendation, especially since changing dietary fat intake was unlikely to have any deleterious effects. They proposed that at least those patients at high risk (e.g. because of extremely high cholesterol levels, or already existing coronary artery disease) should be urged to lower their cholesterol levels by dietary means.

The American Heart Association (AHA), as early as 1961, went on record to recommend reducing dietary fat to no more than 25–35 percent of total calories, reducing total calorie intake, and substituting polyunsaturated fats for saturated fats (65). They guardedly said, "Those people who have had one or more atherosclerotic heart attacks or strokes *may* [emphasis added] reduce the possibility of recurrences by such a change in diet." Time has proved how right they were but at that time the evidence was slim. The AHA dietary guidelines were faulted because direct evidence that lowering *total* fat intake would reduce coronary heart disease risk was not available at the time. What that criticism overlooks is that at the time (and still today) most Americans take in a large amount of saturated fat. Consequently, reducing total fat intake would in almost every case reduce saturated fat intake also and would therefore lower serum cholesterol. Adding polyunsaturated fat to replace saturated fat would help keep the diet acceptable.

A BRIEF WRAP-UP OF THE CASE AGAINST CHOLESTEROL AS OF 1970

At the outset we suggested that in 1970 the "convinced" and the "unconvinced" were really not looking at the same data sets. The "unconvinced" were largely confining themselves to the intervention trial data per se. The "convinced," if they had had a catechism, might have recited it something like this:

1. Cholesterol derived from plasma lipoproteins is a consistent and striking feature of atherosclerotic lesions (66).

2. Lesions similar to those of human atherosclerosis can be produced in many different animal species by raising the level of cholesterol in their blood to a high enough level and maintaining it long enough (67).

3. People with dramatically high blood cholesterol levels, as in familial hyper-cholesterolemia, have dramatically premature coronary heart disease (8–12).

4. People with even relatively modest elevations of blood cholesterol levels are at significantly higher risk. This is true across a wide spectrum of blood cholesterol levels and holds on comparison of populations from different countries (16) and also within populations (19;20).

5. Blood cholesterol level is increased when dietary saturated fat intake is increased, as shown by carefully controlled metabolic ward studies (32;33). Moreover, populations with dietary habits that include a high saturated fat intake have higher blood cholesterol levels and a higher coronary heart disease incidence than populations with lower saturated fat intake (16).

6. The wide differences in blood cholesterol levels and coronary heart disease risk between populations of different countries is largely due to environmental factors (probably diet) rather than genetic factors, as shown, for example, by the Japanese migration studies (18).

7. Dietary intervention to lower blood cholesterol by decreasing saturated fat intake in favor of polyunsaturated fat intake reduces blood cholesterol levels and decreases risk of coronary heart disease and other atherosclerotic complications (45–47;55;56–63).

8. We should be starting to treat hypercholesterolemia in high-risk patients.

But the controversy continued.

THE PROS

In 1964, writing an Editorial about the National Diet–Heart Study, I.H. Page cited a cartoon showing a society matron in her doctor's office asking "Are you a cholesterol doctor or one of those 'cholesterol-is-all-bosh' doctors?" (68). Page and the members of his committee were pioneers, sufficiently persuaded by the evidence available to invest a significant piece of their time and energy in a quest for the definitive clinical trial. They were by no means evangelical. In his Editorial, Page put it this way, "No one knows whether a free-living, healthy population will continue with the changes in eating habits we suggest. Nor do we know whether, if their blood cholesterol levels are moderately reduced, heart attacks will decrease. At least we will try to find out." As pointed out above, the definitive Diet–Heart trial, i.e. a double-blind trial in healthy people, was never done. It would have been impossibly complex and impossibly expensive.

Jermiah Stamler, a member of Page's committee, was a major figure in the battle to gain acceptance of the lipid hypothesis. His early work with Louis N. Katz on atherosclerosis in chickens and their jointly written book *Experimental Atherosclerosis* published in 1953 (69) helped the case for an activist approach. Stamler continued to be an effective spokesman as an epidemiologist of major stature on the world stage throughout his career. Katz, in his 1970 Duff Lecture before the American Heart Association's Council on Arteriosclerosis, urged, "Planned primary and secondary prevention is now already justified in persons with a high degree of proneness to coronary heart disease" (70).

Another early proponent of the lipid hypothesis was William Dock. His 1958 Editorial made the case forcefully (71) and in a 1974 Editorial used the provocative title; "Atherosclerosis: why do we pretend the pathogenesis is mysterious?" (72). When Dock attended meetings he often carried with him a photomicrograph of the coronary artery of a newborn infant showing subintimal lipid deposition.

John Gofman was another who found the Anitschkow rabbit findings persuasive and set out to try to relate them to the human disease. His enormous contribution is dealt with in detail in Chapter 4.

In 1969, the Chairman of the Council on Arteriosclerosis of the American Heart Association proposed that, "It is now good medical practice to treat – and I use the word advisedly – people who have definite hyperlipoproteinemia. In short, we have come ... to the point where we are probably preventing a disease that was considered to be an inevitable accompaniment of aging not very long ago" (73). It would be another 15 years before this point of view would prevail.

THE CONS

There were definitely opposing points of view. Sir John McMichael, a "Dean" of British cardiology, took the gloves off in an editorial essay (74) threateningly entitled "Fats and atheroma: an inquest." He summarized his evaluation of the data available in 1979 this way: "All well-controlled trials of cholesterol-reducing diets and drugs have failed to reduce coronary heart disease mortality and morbidity." Elsewhere he lamented, "some of our profession are stretching so much speculative and insecure evidence to support a dietetic theory no longer [held] tenable by informed medical scientists."

Here in the U.S., George V. Mann, a physician and nutrition expert at Vanderbilt University, dismissed the evidence from the dietary trials as totally unconvincing. In a *New England Journal of Medicine* review (75) entitled "Diet–Heart: end of an era," he suggested that: "The dietary dogma was a money-maker for segments of the food industry, a fund raiser for the Heart Association, and busy work for thousands of fat chemists" and, perhaps plaintively, that: "To be a dissenter was to be unfunded because the peer-review system rewards conformity and excludes criticism."

The cholesterol wars continued apace.

NOTE

1 Here was another large-scale demonstration that removing saturated fat from the diet lowers serum cholesterol significantly.

REFERENCES

1. Lehzen, G. and Knauss, K. 1889. Ueber Xanthoma multiplex planum, tuberosum, mollusciformis. *Virchows Arch fur path Anat* 116:85–104.
2. Pinkus, F. and Pick, L. 1908. Zur Struktur und Genese der symptomatischen Xanthome. *Deutsch med Wchnschr* 34:1426–1430.
3. Fleissig, J. 1913. Uber die bisher als Riesenzellensarkome (Myelome) bezeichneten Granulationsgeschwulste der Sehnenschiden. *Dtsch Z Chir* 122:239–265.
4. Hoessli, H. 1914. Ueber Xanthom der Haut und der Sehnen. *Beitr Klin Chirurg* 90:168–178.
5. Arning, E.L.A. 1920. Essentielle Cholesterinamie mit Xanthomabildung. *Z Klin Med* 89:107–119.
6. Harbitz, F. 1927. Tumors of the tendon sheaths, joint capsules and multiple xanthomas. *Arch Path* 4:507–527.
7. Thannhauser, S.J. and Magendantz, H. 1938. The different clinical groups of xanthomatous diseases: a clinical physiological study of 22 cases. *Ann Int Med* 11:1662–1746.
8. Muller, C. 1939. Angina pectoris in hereditary xanthomatosis. *Arch Int Med* 64:675–700.
9. Wilkinson, C.F., Hand, E.A., and Fliegelman, M.T. 1948. Essential familial hypercholesterolemia. *Ann Intern Med* 29:671–686.

10. Adlersberg, D., Parets, A.D., and Boas, E.P. 1949. Genetics of atherosclerosis. *JAMA* 141:246–254.
11. Khachadurian, A.K. 1964. The inheritance of essential familial hypercholesterolemia. *Am J Med* 37:402–407.
12. McGinley, J., Jones, H., and Gofman, J. 1952. Lipoproteins and xanthomatous diseases. *J Invest Dermat* 19:71–82.
13. Keys, A., Anderson, J.T., Fidanza, F., Keys, M.H., and Swahn, B. 1955. Effects of diet on blood lipids in man, particularly cholesterol and lipoproteins. *Clin Chem* 1:34–52.
14. Keys, A., Aravanis, C., Blackburn, H.W., Van Buchem, F.S., Buzina, R., Djordjevic, B.D., Dontas, A.S., Fidanza, F., Karvonen, M.J., Kimura, N., Lakos, D., Monti, M., Puddu, V., and Taylor, H.L. 1966. Epidemiological studies related to coronary heart disease: characteristics of men aged 40–59 in seven countries. *Acta Med Scand Suppl* 460:1–392.
15. Keys, A. ed. 1970. Coronary disease in seven countries. *Circulation* 41 (Suppl. 1):1–211.
16. Keys, A. 1980. *Seven Countries.* Harvard University Press, Cambridge, MA:1–381.
17. Blackburn, H. 1995. *On the Trail of Heart Attacks in Seven Countries.* The Country Press, Inc., Middleborough, MA:1–148.
18. Robertson, T.L., Kato, H., Rhoads, G.G., Kagan, A., Marmot, M., Syme, S.L., Gordon, T., Worth, R.M., Belsky, J.L., Dock, D.S., Miryanishi, M., and Kawamoto, S. 1977. Epidemiologic studies of coronary heart disease and stroke in Japanese men living in Japan, Hawaii and California. Incidence of myocardial infarction and death from coronary heart disease. *Am J Cardiol* 39:239–243.
19. Kannel, W.B., Dawber, T.R., Kagan, A., Revotskie, N., and Stokes, J., III. 1961. Factors of risk in the development of coronary heart disease – six year follow-up experience. The Framingham Study. *Ann Intern Med* 55:33–50.
20. Wilson, P.W., Garrison, R.J., Castelli, W.P., Feinleib, M., McNamara, P.M., and Kannel, W.B. 1980. Prevalence of coronary heart disease in the Framingham Offspring Study: role of lipoprotein cholesterols. *Am J Cardiol* 46:649–654.
21. Multiple Risk Factor Intervention Trial Research Group. 1982. Multiple risk factor intervention trial. Risk factor changes and mortality results. *JAMA* 248:1465–1477.
22. Frantz, I.D., Jr., Dawson, E.A., Ashman, P.L., Gatewood, L.C., Bartsch, G.E., Kuba, K., and Brewer, E.R. 1989. Test of effect of lipid lowering by diet on cardiovascular risk. The Minnesota Coronary Survey. *Arteriosclerosis* 9:129–135.
23. 1984. The Lipid Research Clinics Coronary Primary Prevention Trial results. I. Reduction in incidence of coronary heart disease. *JAMA* 251:351–364.
24. 1984. The Lipid Research Clinics Coronary Primary Prevention Trial results. II. The relationship of reduction in incidence of coronary heart disease to cholesterol lowering. *JAMA* 251:365–374.
25. Huttunen, J.K., Heinonen, O.P., Manninen, V., Koskinen, P., Hakulinen, T., Teppo, L., Manttari, M., and Frick, M.H. 1994. The Helsinki Heart Study: an 8.5-year safety and mortality follow-up. *J Intern Med* 235:31–39.
26. Page, I.H., Stare, F.J., Corcoran, A.C., Pollack, H., and Wilkinson, C.F., Jr. 1957. Atherosclerosis and the fat content of the diet. *Circulation* 16:163–178.
27. Jacobs, D.R., Jr., Anderson, J.F. and Blackburn, H. 1979. Diet and serum cholesterol: do zero correlations negate the relationship? *Am J Epidemiol* 110:77–87.
28. Blackburn, H. and Jacobs, D. 1984. Sources of the diet–heart controversy: confusion over population versus individual correlations. *Circulation* 70:775–780.
29. DeLangen, C.D. 1916. Cholesterol-metabolism and racial pathology {Dutch}. *Geneeskundig tijdschrift voor Nederlandisch-Indie* 56:1–34.
30. DeLangen, C.D. 1922. Cholesterol contents of blood in the Dutch Indies {Dutch}. *Geneeskundig tijdschroft voor Nederlandisch-Indie* 62:1–4.

31. Groen, J.J., Tjiong, B.K., Kamminga, C.E., and Willebrands, A.F. 1952. The influence of nutrition, individuality and some other factors, including various forms of stress, on the serum cholesterol: an experiment of nine months duration in 60 normal human volunteers. *Voeding* 13:556–587.

32. Kinsell, L.W., Partridge, J., Boling, L., Margen, S., and Michael, G. 1952. Dietary modification of serum cholesterol and phospholipid levels. *J Clin Endocrinol Metab* 12:909–913.

33. Ahrens, E.H., Jr., Blankenhorn, D.H., and Tsaltas, T.T. 1954. Effect on human serum lipids of substituting plant for animal fat in diet. *Proc Soc Exp Biol Med* 86:872–878.

34. Keys, A., Anderson, J.T., and Grande, F. 1957. Prediction of serum-cholesterol responses of man to changes in fats in the diet. *Lancet* 273:959–966.

35. Mayer, G.A., Connell, W.F., Dewolfe, M.S., and Beveridge, J.M. 1954. Diet and plasma cholesterol levels. *Am J Clin Nutr* 2:316–322.

36. Beveridge, J.M., Connell, W.F., and Mayer, G.A. 1956. Dietary factors affecting the level of plasma cholesterol in humans: the role of fat. *Can J Biochem Physiol* 34:441–455.

37. Beveridge, J.M., Connell, W.F., Mayer, G.A., and Haust, H.L. 1960. The response of man to dietary cholesterol. *J Nutr* 71:61–65.

38. Olson, F., Michaels, G., Partridge, J., Boling, L., Margen, S., and Kinsell, L.W. 1953. The use of formula diets administered via polyethylene tube or orally for constant intake (balance) studies. *Am J Clin Nutr* 1:134–139.

39. Ahrens, E.H., Jr., Dole, V.P., and Blankenhorn, D.H. 1954. The use of orally-fed liquid formulas in metabolic studies. *Am J Clin Nutr* 2:336–342.

40. Ahrens, E.H., Jr., Insull, W., Jr., Blomstrand, R., Hirsch, J., Tsaltas, T.T., and Peterson, M.L. 1957. The influence of dietary fats on serum-lipid levels in man. *Lancet* 272:943–953.

41. Beveridge, J.M., Connell, W.J., and Mayer, G.A. 1956. Dietary factors affecting the level of plasma cholesterol in humans: the role of fat. *Can J Med Sci* 34:441–455.

42. Bronte-Stewart, B. and Antonis, A.E.L.B.J.F. 1956. Effects of feeding different fats on serum-cholesterol level. *Lancet* 270:521–527.

43. Keys, A. 1957. Diet and the epidemiology of coronary heart disease. *J Am Med Assoc* 164:1912–1919.

44. Clarke, R., Frost, C., Collins, R., Appleby, P., and Peto, R. 1997. Dietary lipids and blood cholesterol: quantitative meta-analysis of metabolic ward studies. *Br Med J* 314:112–117.

45. Leren, P. 1966. The effect of plasma cholesterol lowering diet in male survivors of myocardial infarction. A controlled clinical trial. *Acta Med Scand Suppl* 466:1–92.

46. Dayton, S., Pearce, M.L., Goldman, H., Harnish, A., Plotkin, D., Shickman, M., Winfield, M., Zager, A., and Dixon, W. 1968. Controlled trial of a diet high in unsaturated fat for prevention of atherosclerotic complications. *Lancet* 2:1060–1062.

47. Miettinen, M., Turpeinen, O., Karvonen, M.J., Elosuo, R., and Paavilainen, E. 1972. Effect of cholesterol-lowering diet on mortality from coronary heart-disease and other causes. A twelve-year clinical trial in men and women. *Lancet* 2:835–838.

48. 1968. The National Diet–Heart Study Final Report. *Circulation* 37:I1–428.

49. Taubes, G. 2001. Nutrition. The soft science of dietary fat. *Science* 291:2536–2545.

50. Posner, B.M., Cobb, J.L., Belanger, A.J., Cupples, L.A., D'Agostino, R.B., and Stokes, J., III. 1991. Dietary lipid predictors of coronary heart disease in men. The Framingham Study. *Arch Intern Med* 151:1181–1187.

51. Nichols, A.B., Ravenscroft, C., Lamphiear, D.E., and Ostrander, L.D., Jr. 1976. Independence of serum lipid levels and dietary habits. The Tecumseh study. *JAMA* 236:1948–1953.

52. Connor, W.E. and Connor, S.L. 2002. Dietary cholesterol and coronary heart disease. *Curr Atheroscler Rep* 4:425–432.

53. Kern, F., Jr. 1991. Normal plasma cholesterol in an 88-year-old man who eats 25 eggs a day. Mechanisms of adaptation. *N Engl J Med* 324:896–899.

54. Weggemans, R.M., Zock, P.L., and Katan, M.B. 2001. Dietary cholesterol from eggs increases the ratio of total cholesterol to high-density lipoprotein cholesterol in humans: a meta-analysis. *Am J Clin Nutr* 73:885–891.

55. Leren, P. 1970. The Oslo diet–heart study. Eleven-year report. *Circulation* 42:935–942.

56. Turpeinen, O. 1968. Diet and coronary events. *J Am Diet Assoc* 52:209–213.

57. Morrison, L.M. 1951. Reduction of mortality rate in coronary atherosclerosis by a low cholesterol, low fat diet. *Am Heart J* 42:538–545.

58. Morrison, L.M. 1955. A nutritional program for prolongation of life in coronary atherosclerosis. *JAMA* 159:1425–1428.

59. Pritikin, N. 1984. The Pritikin diet. *JAMA* 251:1160–1161.

60. Ornish, D., Scherwitz, L.W., Billings, J.H., Brown, S.E., Gould, K.L., Merritt, T.A., Sparler, S., Armstrong, W.T., Ports, T.A., Kirkeeide, R.L., Hogeboom, C., and Brand, R.J. 1998. Intensive lifestyle changes for reversal of coronary heart disease. *JAMA* 280:2001–2007.

61. Christakis, G., Rinzler, S.H., Archer, M., Winslow, G., Jampel, S., Stephenson, J., Friedman, G., Fein, H., Kraus, A., and James, G. 1966. The anti-coronary club. A dietary approach to the prevention of coronary heart disease – a seven-year report. *Am J Public Health Nations Health* 56:299–314.

62. Bierenbaum, M.L., Green, D.P., Florin, A., Fleischman, A.I., and Caldwell, A.B. 1967. Modified-fat dietary management of the young male with coronary disease. A five-year report. *JAMA* 202:1119–1123.

63. 1968. Controlled trial of soya-bean oil in myocardial infarction. *Lancet* 2:693–699.

64. Leren, P. 1968. The effect of plasma-cholesterol-lowering diet in male survivors of myocardial infarction. A controlled clinical trial. *Bull N Y Acad Med* 44:1012–1020.

65. 1965. Diet and the possible prevention of coronary atheroma; a council statement. *JAMA* 194:1149–1150.

66. Windaus, A. 1910. Uber den Gehalt nirmaler und atheromatoser Aorten an Cholsterin und Cholesterinestern. *Hoppe-Seyler Z Physiol Chemie* 67:174–176.

67. Anitschkow, N. 1933. Experimental atherosclerosis in animals. In *Arteriosclerosis*, E.V. Cowdry, ed. Macmillan, New York:271–322.

68. Page, I.H. 1964. The national diet–heart study. *Circulation* 29:4–5.

69. Katz, L.N. and Stamler, J. 1953. *Experimental Atherosclerosis*. Charles C. Thomas, Springfield, Illinois:1–375.

70. Katz, L.N. 1972. Has knowledge of atherosclerosis advanced sufficiently to warrant its application to a practical prevention program? The 1970 G. Lyman Duff Memorial Lecture, Council on Arteriosclerosis, American Heart Association. *Circulation* 45:8–20.

71. Dock, W. 1958. Research in arteriosclerosis; the first fifty years. *Ann Intern Med* 49:699–705.

72. Dock, W. 1974. Atherosclerosis. Why do we pretend the pathogenesis is mysterious? *Circulation* 50:647–649.

73. Steinberg, D. 1970. Progress, prospects and provender. Chairman's address before the Council on Arteriosclerosis, American Heart Association, Dallas, Texas, November 12, 1969. *Circulation* 41:723–728.

74. McMichael, J. 1979. Fats and Atheroma – an Inquest. *Br Med J* 1:173–175.

75. Mann, G.V. 1977. Diet-Heart: end of an era. *N Engl J Med* 297:644–650.

CHAPTER 4

Building the Basic Science Foundation

Advances in medicine almost always rest on a foundation of basic science. The development of tools for testing the lipid hypothesis and for discovering, ultimately, ways to reduce the risk of atherosclerotic complications was no exception. Thus, the progress toward a consensus regarding the pathogenetic pathway for atherosclerosis, discussed in Chapter 5, rested on a wide array of prior advances in basic physiology, cell biology, and molecular biology. Without the elegant studies of Bloch, of Lynen, and of Cornforth and Popjak on the pathway of cholesterol biosynthesis there would have been no statins (1–5). Without the stimulus of the pioneering work of Gofman we might still be stuck with just "α-lipoprotein" and "β-lipoprotein," which is where we were in 1951 (6;7). Actually each of these advances rested in turn on prior basic science advances. Without the discovery of cholesterol in gallstones by Chevreul and the elucidation of its complex structure by Windaus and Wieland, there would have been no way to deduce the pathways of its bio-synthesis and the mechanisms regulating it. Without Urey's discovery of deuterium, Rittenberg and Schoenheimer could not have demonstrated the dynamic state of cholesterol and the fact that it must be synthesized not from a similarly complex molecule but rather from a large number of much smaller molecules – acetate. Space does not permit a detailed examination of all the relevant basic science advances. That would require tracing the history of pathology, pharmacology, biological chemistry, genetics, immunology, cell biology, and molecular biology. All have contributed to our understanding of this very complex disease called atherosclerosis. Discussion will be confined to some key elements that contributed more or less directly to the ultimate acceptance of the lipid hypothesis and to the development of ways to reduce the toll of coronary heart disease.

The Cholesterol Wars by D. Steinberg.
Copyright © 2007 Elsevier Inc. All rights of reproduction in any form reserved.

THE BIOSYNTHESIS OF CHOLESTEROL AND ITS METABOLIC REGULATION

SCHOENHEIMER

Rudolph Schoenheimer's application of stable isotopes to the study of metabolic pathways inaugurated a new era in biochemistry (8). Schoenheimer, born in Berlin, was teaching at the University of Freiburg in Ludwig Aschoff's Department of Pathology when the Nazis came to power. Jews were soon forbidden to hold university appointments and Schoenheimer was forced to emigrate. Hans Clarke, Chair of the Department of Biochemistry at Columbia University, a man sympathetic to the plight of refugee scientists, offered him a position in 1933. Just a year earlier, Harold Urey, in the Department of Chemistry at Columbia, had discovered deuterium, a discovery that won him a Nobel Prize in 1934. Schoenheimer attended a seminar by Urey and saw immediately the enormous potential of an isotope of hydrogen in the elucidation of biochemical transformations. His mind was probably fertile soil for this new approach to biochemistry because of his earlier encounter with George Hevesy during the latter's visit to Freiburg in 1930 (9). Hevesy, a physicist, had shown how a naturally occurring radioactive isotope of lead could be used to trace movements of minute amounts of lead from tissue to tissue, probably the first application of isotopes in biology. Hevesy, a physical chemist by training, had come to Aschoff's laboratory to learn more about biology. Schoenheimer was, at the time, an assistant to Aschoff and he drew the assignment to collaborate with Hevesy. So Schoenheimer must have come to Urey's lecture with a well-prepared mind. He saw immediately the power of a tool that would allow fluxes to be measured without perturbing the steady state of biological systems (10). At that time, research on metabolic pathways pretty much depended on the balance method: feed or add to the biological system under study an amount of substance "A"; if the amount of substance "B" increases, then possibly "A" is a metabolic precursor of "B" but not necessarily. The increase in "B" could result from metabolic shifts induced by "A" rather than direct interconversion. As Schoenheimer liked to point out, introducing a coin into a Coca-Cola dispenser produces a bottle of Coke but does not prove that the coin has been converted to a bottle of Coke. With deuterium and ^{15}N, generously provided by Harold Urey, Schoenheimer and his colleagues in that remarkable Department of Biochemistry at Columbia showed that almost all body constituents are in a dynamic state, constantly undergoing degradation at some rate and being resynthesized at a comparable rate to maintain steady state (Figure 4.1).

With David Rittenberg, Schoenheimer showed, in 1937, that this was the case for cholesterol (11). Rittenberg had recently completed his Ph.D. thesis in physical chemistry under Urey's direction and brought with him to Schoenheimer's

FIGURE 4.1 Rudolf Schoenheimer in his laboratory in the Department of Biochemistry at Columbia University, College of Physicians and Surgeons, *ca.* 1937. Source: reprinted from reference (9) with permission.

laboratory all of the techniques for quantification of deuterium. Together they injected mice with enough concentrated D_2O to make the deuterium content about 1.5 percent and then fed enough D_2O daily to maintain that same level of enrichment. After 3 months the hydrogen in the cholesterol isolated from the carcasses showed an enrichment almost half that of the body water. This level of incorporation was so striking that they had to conclude that cholesterol must be built up from many small molecules, "possibly those which have been postulated to be intermediates in the fat and carbohydrate metabolism."

That same year, Sonderhoff and Thomas, studying the nonsaponifiable lipids in yeast, reported a very high level of incorporation of trideuteroacetate and concluded that the sterols must be synthesized in a direct fashion from acetate (12). Prior to this work, speculations about the origin of body cholesterol had tended to focus on possible large molecule precursors; the new findings ruled out any such pathways because they would not entail the large incorporation of deuterium from D_2O during transformation. Acetate must be a major precursor but was it the only precursor? And how could such a simple

2-carbon molecule be the major building block for the complex 27-carbon, 4-ringed structure of the cholesterol molecule?

BLOCH, LYNEN, AND THE CORNFORTH/POPJAK TEAM

In the late 1930s, another young Jewish émigré from Germany, Konrad Bloch, joined Clarke's department as a graduate student. Bloch had already completed most of his thesis research at the University of Basel and had published two papers on that research (Figure 4.2). Still, the Basel faculty rejected it as "insufficient" (10). Bloch many years later learned that only one examiner on his committee had objected and that was on the grounds that the thesis failed to cite some important references – papers authored by that examiner! Looking back, Bloch realized that this may have been providential. Had he passed he may have decided to stay on in Germany.

At any rate, when Bloch came to New York in 1936, Clarke, a guardian angel to refugee scientists, admitted him to his program and the Ph.D. was awarded about 2 years later. At that point Schoenheimer offered Bloch a position in his

FIGURE 4.2   Konrad Bloch, winner of the Nobel Prize in Physiology or Medicine in 1964 shared with Feodor Lynen. Source: reprinted from reference (9) with permission.

laboratory and Bloch became intrigued by the power of the isotope approach in general and its application to cholesterol metabolism in particular. In 1941, Schoenheimer, at age 43, tragically took his own life for reasons not known. Needless to say the laboratory was thrown into turmoil, but Clarke encouraged the young people to stay on and exploit the methodology that Schoenheimer had pioneered. Bloch recalls that, whether by drawing lots or in some other way, David Shemin "drew" amino acid metabolism; David Rittenberg got protein synthesis; and, Bloch says, "lipids were to be my territory" (10). He ploughed that territory with diligence and ingenuity for the next 20 years, winning the Nobel Prize in 1965.

While still at Columbia, Bloch collaborated with Rittenberg to show that acetate contributes in a major way to both the side chain and the ring structure of cholesterol (13). Later, at the University of Chicago, Bloch began to examine the sources of individual carbon atoms in the cholesterol molecule and predicted that *all* would turn out to be derived from acetate (14). Proof came from a later collaboration with E.L. Tatum, who had isolated a strain of *Neurospora crassa* that could not metabolize pyruvate and was therefore totally dependent on an exogenous source of acetate for growth. When grown in the presence of labeled acetate, the sterol of the organism, ergosterol, showed essentially no dilution of the isotope (15).

Now the challenge was to figure out the predictably complex steps in the conversion of acetate to cholesterol. Bloch's thinking was greatly influenced by the work of Bonner and Arreguin on how plants synthesize rubber, an isoprene polymer (16). They showed that acetate was utilized to generate isoprene subunits and proposed a scheme by which three acetate molecules could give rise to an isoprenoid subunit. If an analogous pathway operated in mammalian systems one could predict which carbon atoms in the isooctyl side chain of cholesterol would be derived from the methyl carbon and which from the carboxyl carbon of acetate. In 1952, Bloch's laboratory experimentally confirmed the predicted pattern. This also agreed with the pattern expected if, as Robinson had postulated as early as 1934, cholesterol was formed by cyclization of squalene (17). The following year, Langdon and Bloch showed directly that labeled squalene could indeed be converted to cholesterol by the rat (18;19). Without doubt it was Bloch who led the way in the elucidation of the pathway for cholesterol biosynthesis.

By 1952, everyone was convinced that three acetate units must somehow give rise to a six-carbon intermediate, which must then lose one carbon to generate a five-carbon isopentenyl precursor. However, the nature of that intermediate proved elusive. The breakthrough came serendipitously when a group at Merck, while looking for nutritive materials that could substitute for acetate in an acetate-dependent bacterium, isolated and characterized for the first time mevalonic acid. Noting the close similarity in structure between mevalonate and hydroxymethylglutarate, they showed that the former was almost quantitatively converted to squalene and sterols (20). The pathway to mevalonate via acetoacetate and hydroxymethylglutarylCoA was quickly demonstrated by Rudney (21) and by Lynen's group (2). The exciting discovery of the pathway from mevalonate to squalene via phosphorylated isopentenyl derivatives is reviewed elsewhere in detail (1;3;10;22;23).

Robinson's speculation about the cyclization of squalene, while important heuristically, turned out to be not quite correct. The publication of the correct structure of lanosterol in Ruzicka's laboratory in Zurich in 1952 (24), suggested to R.B. Woodward at Harvard an alternative way to fold squalene that would fit exactly the structure of lanosterol and, by inference, put lanosterol on the pathway from squalene to cholesterol. Bloch was visiting Woodward at Harvard at the time and was persuaded by the logic of the proposal. He saw immediately that the origin of the carbon at position 13 would critically differentiate the Robinson and the Woodward proposals. He returned to his laboratory, isolated that carbon atom, and found that it fitted with the Woodward proposal and not with the Robinson proposal (25).

Over the next several years the origins (methyl versus carboxyl) of all 27 carbon atoms in cholesterol were established in a series of elegant studies from Bloch's laboratory and from the team of G. Popjak and J.W. Cornforth at the British National Institute for Medical Research at Mill Hill outside London (reviewed in 1;23;5).

John Cornforth was a brilliant organic chemist who had decided to tackle the daunting problem of a *de novo* total synthesis of the cholesterol molecule (Figure 4.3). Cornforth began to lose his hearing when he was only 10 and by age 20 he was totally deaf, but he was an amazing lip reader. You just had to remember to look directly at him when you spoke. He was one of those chemists who seemed to have an instinctive feel for the right approach to any problem in organic chemistry you might consult him on. He was also a formidable chess player, competing for the New South Wales Chess Championship in 1937 at age 20. George Popjak, a Hungarian expatriate, who was also working at Mill Hill at the time, proposed to Cornforth that they join hands in a combined chemical and biochemical attack on the cholesterol biosynthesis problem: that was the beginning of a 20-year collaboration interrupted only when Popjak decided to leave England for a post at the University of California, Los Angeles in 1968. Cornforth was arguably the most outstanding researcher of his time working on the chemistry of natural products. He was an absolute artist in the area of stereochemistry and the biosynthesis of stereospecific products. His Nobel Prize in Chemistry in 1975, shared with Vladimir Prelog, was awarded "for his work on the stereochemistry of enzyme-catalyzed reactions."

Cornforth's work on cholesterol with Popjak was one facet, and an important facet, of his claim to the Prize, but it was the combination of that with his prior

FIGURE 4.3   John Warcup Cornforth, winner of the Nobel Prize in Chemistry in 1975, shared with Vladimir Prelog. Source: reprinted with permission; Copyright The Nobel Foundation.

studies and his post-Popjak studies that made the case. The prize for cholesterol biosynthesis had already been awarded to Bloch and Lynen in 1965. Some felt that Popjak had "fallen between two prizes." However, Popjak's enormous contributions did not by any means go unrecognized (26). He was awarded the Ciba Medal, the Stouffer Prize, the Royal Society's Davy Medal, and the Award in Lipid Chemistry from the American Oil Chemists Society (Figure 4.4).

From the basic work in these several laboratories emerged a completed canvas detailing the steps for the biosynthesis of the cholesterol molecule. How would that be used in connection with atherosclerosis?

THE RATE-LIMITING STEP, HMGCoA REDUCTASE

The possibility that inhibition of cholesterol synthesis might lower blood cholesterol levels was one of the first things to come to mind. As discussed in

FIGURE 4.4   George Joseph Popjak, whose long-standing collaboration with J.W. Cornforth played a critical role in the elucidation of the pathway for cholesterol biosynthesis. Source: reprinted from reference (26) with permission.

Chapter 8 many inhibitors were tested but they were either ineffective *in vivo* or toxic (27). The first clue as to the nature of the rate-limiting step came from R.G. Gould's demonstration that cholesterol itself exerted feedback inhibition on cholesterol synthesis, reported in abstract form by Gould and Taylor in 1950 (28), and in more extended form a few years later (29). Gould later collaborated with Popjak to show directly that the rate-limiting step was the conversion of hydroxymethylglutarate to mevalonate, i.e. the HMGCoA reductase step (30). This then became the logical target for inhibitors that might lower blood cholesterol levels. Many structural analogues were synthesized and tested but with uniformly disappointing results. In fact an effective and safe inhibitor did not surface until 1976, when Endo discovered the first of the statin drugs (31). Further discussion of this is deferred to Chapter 8.

THE BIRTH OF THE LIPOPROTEINS AND THE JOHN W. GOFMAN STORY

During the first two decades of the 20th century, a number of investigators drew attention to the fact that the considerable amounts of lipid in serum must be present either in some sort of emulsion or in association with proteins. The concentration of lipids in plasma far exceeded – by orders of magnitude – their solubility in aqueous media. However, the first definitive studies on the nature of the plasma lipoproteins were reported by Macheboeuf beginning in 1929 (32). Over the next 10 or 20 years he succeeded in purifying and carefully characterizing α-lipoprotein from horse serum and showed that the lipid-free protein remained soluble. During the Second World War, E.J. Cohn and J.L. Oncley and their collaborators at Harvard developed elaborate large-scale methods for fractionating human serum to provide materials useful in treating the wounded. In the course of those systematic studies they found that the lipids of serum were concentrated in two major fractions having α_1- and β-mobility, respectively (6;33).

At the time nothing was known about their origin, their fate, or their biological significance. Then, in 1951, E.M. Russ, H.A. Eder, and D.P. Barr at the New York Hospital-Cornell Medical Center, using the methods developed in the Cohn/Oncley laboratory, made the important finding that women prior to menopause had consistently higher blood levels of the α-lipoprotein than did men (34). They speculated that this difference might be related to the lower incidence of coronary artery disease in premenopausal women – and time has certainly proved them right. Their 1951 paper was possibly the first demonstration that distinct lipoprotein classes might have distinct biological functions. It also was the first suggestion of a linkage between specific lipoprotein patterns and the risk of coronary heart disease. However, their speculation was at that time based only on a correlation and did not establish a *causal* relationship. Today that causal relationship has been established beyond a doubt.

John W. Gofman was not the first to try characterizing the full spectrum of lipoproteins in the blood, but he was the first to do so successfully. Gofman is a prime example of the unusual man who straddles two fields and as a result is able to see novel ways of applying methods and ideas from one field to the other. He started medical school at Case Western University in Cleveland but before finishing went on to the University of California, Berkeley. There he came under the influence of two giants, both Nobel Prize winners – Ernest O. Lawrence, inventor of the cyclotron, and Glenn T. Seaborg, who created ten transuranium elements using that cyclotron. Gofman was quickly swept up in the excitement of the Lawrence laboratory as part of the atomic bomb team. He stayed on and earned a Ph.D. in physics under Seaborg's direction and then entered medical school at the University of California, San Francisco across the bay (Figure 4.5).

Gofman had always been certain he wanted to do biomedical research, and he leaned toward research on cardiovascular disease. He was familiar with Anitschkow's work and he, unlike most others at the time, took it very seriously. Anitschkow's work, together with the genetic, biochemical, and epidemiological evidence available (albeit limited) convinced Gofman that blood cholesterol, and the dietary determinants of blood cholesterol, were centrally important in

FIGURE 4.5   John W. Gofman, whose pioneering work at the Donner Laboratories opened up a new era in research on lipoproteins and their role in atherogenesis. Source: reprinted from reference (46) with permission.

atherosclerosis. His level of conviction is attested to by the Introduction he wrote for a book that his wife, Helen F. Gofman, published in 1951 with several others at Berkeley (35). This was possibly the first low-fat, low-cholesterol "Diet–Heart" cookbook ever published. Clearly Gofman accepted Anitschkow's dictum that cholesterol in the blood somehow played a causative role in atherosclerosis. At the same time, he recognized that almost nothing was known about the chemical form or forms in which that blood cholesterol was carried.

Gofman's unique background enabled him to open up new territory. Because of his strong background in physics and chemistry, he could see the potential power of a then new and highly sophisticated technique – analytic ultracentrifugation. The technique had been developed in Scandinavia by Svedberg and proved to be invaluable for characterizing proteins and measuring their molecular sizes and relative concentrations in mixtures. Gofman wanted to see if it could be used to characterize further the lipoproteins in human serum. At the time, there were only a few such instruments in the entire world. Moreover, Pedersen, the world's expert in Sweden, had already tried to study the proteins in whole serum but had encountered an artifact he could not explain (36). All serum samples contained what he called the "X-protein," present in varying concentrations from sample to sample and, much worse, seeming to change in concentration during the analytical centrifuge run. Pedersen had good evidence, from his own work and that of others, that the X-protein must be a lipoprotein (36–38). He had shown that it floated to the top of the centrifuge tube if salt was first added to the serum and he knew that it was rich in associated phosphatides. After extraction of the lipids (36;37) or after treatment with lecithinase (38), the X-protein disappeared. Pedersen had even estimated its molecular weight at 1.9×10^6, an excellent estimate for LDL! However, he did not recognize the reasons for the apparent changes in concentration during analysis, still thinking they must reflect some reversible aggregation of protein X with other plasma proteins. From the Schlieren pattern he estimated that protein X would have to account for as much as 30 percent of total serum proteins, which seemed implausible. He had given up trying to work with whole serum.

The true explanation of the X-protein artifact was discovered by Gofman with his collaborators, Frank T. Lindgren and Harold Elliot, and published in 1949 (39). Gofman showed very elegantly that the apparent artifact was due to the presence of the very thing he wanted to study – lipoproteins. What was happening was that LDL, less dense than albumin but dense enough to sediment at the density of serum, was also moving down the tube but more slowly. As the concentration of albumin built up, the combined background density of serum and that of the albumin at the boundary now exceeded the density of LDL. Consequently LDL at that point in the tube ceased sedimenting and tended to migrate toward the top of the tube. The Gofman team showed unequivocally that LDL and other lipoproteins were the basis for the protein-X artifact. When they simply added salt to the serum to increase its density so that the lipoproteins floated instead of sedimenting, they got highly reproducible results. Instead of trying to analyze whole serum, they first floated all the lipoproteins, thereby concentrating them and eliminating the X-protein artifact. They went on to devise accurate and reproducible ways to separate the lipoproteins in plasma into subclasses and to measure their concentrations reliably. This was the breakthrough they had been looking for; now they were off and running with a view to studying the relationship between atherosclerosis and the concentrations of the different classes of cholesterol-carrying lipoproteins.

In a 5- or 6-year period beginning in 1949, Gofman and his collaborators turned out a prodigious amount of new information about the lipoproteins in human plasma, their metabolism, and their correlation with atherosclerosis. Their basic "credo" was that it mattered a great deal in which lipoprotein fractions blood cholesterol was carried and, in 1950, they presented preliminary data on a limited number of patients suggesting that the S_f 10–20 fraction was particularly proatherogenic (40). This fraction corresponds closely to IDL (intermediate density lipoproteins), small remnants, which do indeed seem to be strongly proatherogenic. Later they would propose a formula weighting the different subclasses according to their presumed atherogenicity, a so-called "atherogenic index" (41). Whether this index was more predictive than just total blood cholesterol became a highly controversial issue, as discussed below. However, this aside, Gofman's laboratory made a number of important observations and, most important, they started people thinking about the serum lipoproteins and looking into their relationship to atherosclerosis. For example, they identified a small group of patients having characteristic skin xanthomas (xanthoma tuberosum) and having a unique lipoprotein pattern. They suggested that this was a distinct disease entity, one carrying a very high risk of coronary heart disease. What they described is what we now know as dysbetalipoproteinemia, which is indeed the result of a specific mutation in apoprotein E. Gofman and his group also did a pioneering study of the post-heparin clearing phenomenon. They used the analytical ultracentrifuge to demonstrate that the clearing was accompanied by a progressive decrease in the size of the larger lipoproteins to smaller, less buoyant forms, a shift that decreases light scattering (42). That work anticipated the discovery that the whole process was enzymatically triggered by lipoprotein lipase, a discovery made soon thereafter by C.B. Anfinsen, E. Korn, and colleagues at the National Heart Institute, as will be discussed in more detail below.

Gofman's early work showed a good correlation between lipoprotein elevation and risk of heart attack but the sample sizes in those preliminary studies were small. Still, the data suggested that the different classes of lipoproteins should be weighted differently in estimating risk of coronary heart disease, using the "Atherogenic Index" (41). Gofman proposed that this index, or a similar profile of the *kinds* of lipoproteins that were elevated and their concentrations, should predict coronary risk better than just the total cholesterol level. What was needed to clinch the case was a large prospective study. Why didn't Gofman or others interested in the heart disease problem just plow ahead and do such a study? To do so would have required making analytical centrifuge runs on hundreds of individuals. There were at the time only two such instruments in the whole country, they were extraordinarily expensive and technically difficult to operate; finally, most research workers were quite skeptical about the value of what Gofman was doing. Nevertheless, his preliminary results were so impressive that the National Institutes of Health eventually decided to fund a large cooperative project to test Gofman's ideas in a prospective fashion. Gofman tells an interesting story about how that study got off the ground.

Gofman had already applied to the National Institutes of Health for a large grant that would let him move ahead more rapidly. To run the requisite number of samples in a reasonable length of time meant buying an additional analytic centrifuge and expanding the laboratory staff, so his request was for $70,000 per year (equivalent to over $500,000 in current dollars). The application was turned down. Gofman was distraught. Without additional funding he could not critically test the generality of his hypothesis. At about the same time his grant was turned down, he was invited by an old friend, Lawrence Spivak, to write an article on his lipoprotein work for the *American Mercury* magazine, which Spivak edited. (Later Spivak would become more widely known for launching "Meet the Press" with Martha Rountree.)

When Gofman told Spivak what had happened, Spivak promptly placed a telephone call to Mary Lasker to ask for her advice. Mary Lasker was a force to be reckoned with. She was a tireless supporter of biomedical research. She and her husband gave generously of their own money, but, even more important, Mary Lasker directly lobbied the Congress on behalf of NIH. She is credited with playing a role second in importance only to that of James A. Shannon for building the research budget of NIH at an astonishing rate during those early post-war years. Mary Lasker immediately made an appointment to see Gofman at her Beekman Place apartment in New York. She was taken with Gofman and impressed by the promise of what he was doing and she, in turn, made a key telephone call. She called T. Duckett Jones, Professor of Medicine at Harvard, and asked him to come down to New York and meet Gofman. Jones was there the very next day; he listened to Gofman's story and said, "John, you need help. You're not going to get that grant unless you get help. What I think you need to do is to get two or three additional laboratories to agree to join you as collaborators in this study in order to make it saleable. We'll just have to call in some other people."

The upshot of all this was that three other research centers (the Cleveland Clinic, the University of Pittsburgh, and the Harvard School of Public Health) joined hands with Gofman and proposed to the NIH a Cooperative Study using the Gofman type of analysis. So, the NIH, having turned down Gofman's request for $70,000, was now investing $280,000 in a four-center project. The decision may have involved some bending of standard NIH operating procedures, but it turned out to have been a sound one. Over a period of 3 years, the 4 centers analyzed the lipoproteins of almost 5,000 men aged 40–59 who were clinically normal at the time they were first studied. These men were carefully followed over a 3-year period, during which there were 82 cardiac events (myocardial infarction or development of new angina pectoris) classified as definite or probable. The issue was whether the pattern of lipoproteins determined with the analytical centrifuge would or would not be a better predictor of those who were going to have an event than simply the measurement of the total cholesterol in the blood. After the study had already begun, Gofman became aware of a technical glitch in his methodology but also figured out a way to correct it. However, by that time a large number of samples had already been used up or there was not enough left to reanalyze. As a result, the final report from three of the four participating laboratories had to be based only on results obtained using the original, possibly inaccurate method; the results from the Donner Laboratory were corrected and reported in both the original form and the revised, presumably more accurate form. The final report, published in 1956, contained two formal Discussion sections, one representing the views of the Donner Laboratory and the other representing the views of the other three centers, an unfortunate schism (43). However, the results contained important lessons. It was clear that *either* the total cholesterol level *or* the lipoprotein pattern identified those at high coronary heart disease risk. The lipoprotein pattern was not necessarily superior to the measure of total cholesterol level, but it was just as good.

The original protocol for the Cooperative Study defined "definite events" to include angina pectoris, a subjective finding. In Gofman's dissenting report, he analyzed the data with and without angina pectoris included and found that the predictive value of lipoprotein analysis (both total cholesterol and ultracentrifuge analysis) was much greater when this subjective end point was eliminated. At the time there was quite a fuss about the Gofman dissenting report and feelings ran high. What was lost sight of at the time, and even in retrospect, is that in 1956 these investigators had provided important additional evidence that cholesterol-carrying molecules in the blood predicted risk of heart disease. Later studies would show that different lipoprotein fractions do indeed have different degrees of relevance to atherosclerosis, i.e. the phenotype does count. Today, LDL is recognized as the most atherogenic of the lipoproteins, which agrees with Gofman's findings. Later studies also showed that the very low density lipoproteins (VLDL) are less predictive than is LDL but do correlate positively with risk. So, Gofman was basically right but, unfortunately, the data from the Cooperative Study by themselves did not make the case. The analytic ultracentrifuge soon gave way to the preparative ultracentrifuge in lipoprotein research (44) and to paper electrophoresis in clinical research (45). Gofman himself began to turn his attention more and more to the issue of radiation hazards (46) but the Donner Laboratory under Frank T. Lindgren, Alex V. Nichols, and their collaborators continued to exploit the ultracentrifuge as a valuable research tool.

The impact of Gofman's work on the field was of great and lasting importance. He opened the window on the complexity of the lipoproteins and started people thinking about what they do, how they are metabolized, and how they lead to atherosclerosis.

UNRAVELING THE COMPLEX METABOLISM AND INTERACTIONS OF THE PLASMA LIPOPROTEINS

Generating a detailed lipid hypothesis would first require an understanding of the structure and metabolism of the lipoproteins that transported lipids. Over the next several decades there was a ferment of activity in many laboratories around the world that spelled out in detail the metabolic origins and metabolic fates of the lipoproteins, their transport functions, their complex interactions and transformations in the plasma, and the functional importance of the various apolipoproteins as co-factors and as the means of "addressing" lipoprotein particles to their target tissues.

It would require more space than is available here to do justice to this important chapter in atherosclerosis research. An informative and engaging historical review by Donald S. Fredrickson is warmly recommended (47), as are a number of additional reviews by some of the people who developed the field: Richard J. Havel's laboratory (48); Robert W. Mahley's laboratory (49); H. Bryan Brewer's laboratory (50, 52); and Petar Alaupovic's laboratory (51). Here we will limit ourselves to the question of how progress in the lipoprotein field contributed directly to understanding atherogenesis and gaining acceptance of the lipid hypothesis. It did so in several ways:

1. it made the blood cholesterol–coronary heart disease connection concrete at the biochemical and, ultimately, at the molecular level;
2. it provided the basic science substratum on which atherosclerosis research was going to be built; and
3. it provided the building blocks for the development of therapies that would one day make it possible to correct hypercholesterolemia and reduce CHD risk, thus establishing the validity of the lipid hypothesis once and for all.

In the 1940s and 1950s, atherosclerosis research was mostly the province of the pathologists and it was largely descriptive. Very few basic scientists were attracted to the field, in part because they could see no obvious "hooks" on which to hang their biochemical hats. There were plenty of exciting questions in better-plowed fields. The opening up of the lipoprotein field provided the "hook." Here was a dynamic, complicated system for lipid transport that was being explored for the first time. More sophisticated tools for studying lipoproteins in experimental animals and also in humans were becoming available. If blood cholesterol played a role in atherosclerosis, then unlocking the mysteries of the complexes that carried it could be very important. Even if the lipoproteins turned out to be unimportant in atherogenesis, an understanding of them would probably yield important general insights into other aspects of normal and abnormal lipid transport.

WHICH LIPOPROTEINS ARE PROATHEROGENIC?

CHYLOMICRONS?

Because they are so large and get into the arterial wall much more slowly even than VLDL, chylomicrons would appear not to pose much of a threat. And indeed this appears to be the case as long as the chylomicrons are not being broken down to smaller particles at any significant rate, as in the case of patients with familial lipoprotein lipase deficiency or in the case of cholesterol-fed diabetic rabbits. Compared to nondiabetic animals with equal elevations of total blood cholesterol, diabetic cholesterol-fed rabbits, paradoxically, have much *less* atherosclerosis (53). As shown by Zilversmit and colleagues, this can be nicely explained by the fact that chylomicrons and VLDL are so large that they are virtually excluded from the subendothelial space. On the other hand, if lipoprotein lipase is active and the chylomicrons are degraded to smaller, so-called remnant particles, they now can and do enter the artery wall and are decidedly proatherogenic (54;55). The rate of entry of lipoproteins into the artery wall of the rabbit decreases linearly with the logarithm of their molecular diameter (56).

TRIGLYCERIDE-RICH APOB PARTICLES?

The triglyceride level in plasma, considered as an isolated variable, correlates positively with coronary heart disease risk. Whether it is directly involved as a causative factor or is only a marker or "innocent bystander" is difficult to ascertain. For example, obesity is often accompanied by high triglycerides but also by low HDL, diabetes, and hypertension. After appropriate statistical correction for such confounders, plasma triglycerides appear not to be an *independent* risk factor. However, this does not rule out the presence of subsets of cases of hypertriglyceridemia in which the correlation is indeed indicative of cause-and-effect. An example is that group of patients with apoE deficiency and the accompanying accumulation of triglyceride-rich remnants (49;57). When chylomicrons or large triglyceride-rich VLDL are acted on by lipoprotein lipase, with removal of most of the triglyceride, they retain their full complement of apoB and show a high content of apoE. These remnant particles are rich in both cholesterol and triglycerides and float at the density of VLDL but have β electrophoretic mobility (β-VLDL). These remnant particles account for a major part of the hypercholesterolemia in patients with dyslipoproteinemia associated with the apoprotein E2 phenotype because this isoform of apoE binds very poorly to hepatic receptors. These patients are indeed at high risk for atherosclerotic complications, which shows that these are decidedly proatherogenic lipoproteins. Beta-VLDL can be avidly taken up by macrophages even without prior modification, probably by way of the LDL receptor rather than scavenger receptors (57). Whether larger triglyceride-rich lipoproteins are ever proatherogenic and to what extent remains uncertain but, in persons at high risk, more aggressive attempts to control triglyceride levels are now recommended (58).

LOW DENSITY LIPOPROTEIN?

LDL is clearly the most important proatherogenic lipoprotein. It is the major cholesterol-carrying lipoprotein. It is the only fraction markedly elevated in familial
hypercholesterolemia. It is by all odds the lipoprotein fraction most commonly
elevated in the garden-variety hypercholesterolemic patient at high risk for coronary heart disease. Therapy today is guided primarily by the response of LDL
levels to intervention (58).

HIGH DENSITY LIPOPROTEINS?

HDL is of course actually *anti-atherogenic*, as first postulated by Russ, Barr, and
Eder based on the differences between HDL levels in men and in premenopausal
women (7). The fact that low HDL was indeed a negative risk factor was firmly
established in a masterful analysis by Miller and Miller (59). The epidemiological
correlations were clear-cut: a low HDL cholesterol level was associated with a distinctly higher risk for coronary heart disease. However, the mechanisms involved
remained unclear for a long time. As discussed in Chapter 5, later studies led to
the elucidation of the reverse cholesterol transport pathway and the crucial role of
HDL in that process. Recent studies showing that HDL may be atheroprotective
by virtue of its anti-inflammatory properties are also discussed.

THE NATIONAL HEART INSTITUTE STORY

At almost exactly the same time that Gofman was beginning his lipoprotein studies in Berkeley, the NIH, in 1948, established the National Heart Institute, with
James A. Shannon as Associate Director for Research (Figure 4.6). Shannon, who
had trained at New York University under the great guru of kidney physiology,
Homer Smith, was widely known and respected for his personal scientific contributions and for his good judgment and wise leadership when he was Director of
the Squibb Institute for Medical Research. It was now his job to recruit the best
and brightest from their university ivory towers to this new federal laboratory in
Bethesda, MD just outside Washington, DC. This was not an easy job in 1949. At
that time the NIH was not well known. It consisted of just three small buildings
plus the Cancer Institute (the *only* Institute at the time). Although there were
ambitious plans for a huge new research building (which opened as the Clinical
Center in 1954) and a great deal of enthusiasm in the Congress, the words "gov-
ernment laboratory" still had a musty connotation. Career-conscious young acade-
micians instinctively felt that their futures were more assured in academia than in

FIGURE 4.6   James A. Shannon, first Scientific Director of the National Heart Institute and later Director of the National Institutes of Health, who shaped the strong basic research base of the Institutes by his judicious selection of the research staff. Source: courtesy of the National Institutes of Health.

the U.S. Civil Service or in the U.S. Public Health Service. On the other hand, Shannon had some powerful weapons going for him – more research money, better space (once the Clinical Center was finished), his own impeccable credentials as a scientific leader, and his own special brand of Irish charm. Importantly, he could offer *draft deferments*. With those weapons he was successful in putting together a remarkable cadre of talented researchers to head up the divisions of the new Institute, including, among others, E.R. Stadtman, R.W. Berliner, R. Bowman, B. Brodie, S. Sarnoff, E. Horning, B. Witkop, A.G. Morrow – and C.B. Anfinsen.

Anfinsen was a young Assistant Professor of Biological Chemistry at Harvard Medical School (Figure 4.7), who came to Shannon with all-out recommendations from the department chair, A. Baird Hastings. Anfinsen's research interests were in basic enzymology and protein chemistry. Any connection between his research and heart disease was, to say the least, remote. However, Shannon assured Anfinsen that research with a direct linkage to heart disease was not a requirement in his new Institute. The Institute would support research that spanned the full spectrum – from the most basic (Anfinsen's Laboratory of

FIGURE 4.7 Christian B. Anfinsen, who put together a strong group of young investigators with whom he did important pioneering studies of the plasma lipoproteins and made the National Heart Institute one of the leading centers in the field. He shared the Nobel Prize in Chemistry in 1972 with Stanford Moore and William Stein for his work on the folding of ribonuclease. Source: reprinted with permission; Copyright The Nobel Foundation.

Cellular Physiology and Metabolism) to the most applied (Glen Morrow's Cardiac Surgery Branch). Shannon promised Anfinsen that no one would pressure him to move away from the basic questions that interested him. Indeed, that pledge was honored. From the beginning, Anfinsen focused his energies on basic studies of the structure of ribonuclease. He ultimately showed that its catalytic activity depended on its precise configuration or folding, which in turn was determined by the amino acid sequence and proper disulfide bonding. For that work he shared the 1972 Nobel Prize in Chemistry with William H. Stein and Stanford Moore of the Rockefeller University. Clearly, neither Shannon nor Shannon's successors interfered with Anfinsen's basic research. That having been made clear, I think I can now recount what may have represented a minor departure from this strict hands-off policy, a departure with a happy outcome.

In 1950 or 1951, Shannon dropped by Anfinsen's laboratory for a chat. He asked Anfinsen if he was aware of the work being done in California on lipoproteins and their possible relevance to heart disease, the work of a young investigator named Gofman. No, Anfinsen was not aware of it. Shannon said the work was getting a lot of attention and that some members of the Congress had asked him about it. Would Anfinsen look into it and advise Shannon? After all, he said, we are the National *Heart* Institute and we should at least keep track of what's going on. Anfinsen did look into it and, probably to his surprise, found something intriguing, something he thought would be fun to follow up, namely, the nature of what was then called the "clearing factor." The upshot was that Shannon assigned more positions and more laboratory space to Anfinsen and brought the NIH squarely into the lipoprotein era. The "clearing factor" story is a nice example of serendipity in science and it is worth taking a step back to recall it.

In 1943, P.F. Hahn had reported a chance observation he had made while studying factors regulating the mass of red blood cells in dogs (60). The animals were supposed to be fasted overnight before they were studied, but one evening the technician forgot to remove the food from the cages. Hahn's protocol called for drawing a blood sample at the beginning of the experiment, then giving the dog an injection of heparin to prevent clotting, and finally drawing a second blood sample. On this particular day the first samples were visibly cloudy but the second samples, taken just 5 or 10 minutes later, were perfectly clear. Hahn showed that adding heparin directly to the cloudy plasma did not have any clearing effect. The injected heparin must be somehow eliciting the formation of a "clearing factor" in the dog's body. This was evident because just adding some of the clear plasma drawn from a control dog *after* a heparin injection to a sample of cloudy plasma from a nonfasted dog caused it to "clear." Anderson and Fawcett confirmed these observations (61) and speculated that the clearing was due to some physico-chemical disruption induced by heparin-phospholipid complexes. D.M. Graham, in Gofman's laboratory, used the analytic ultra-centrifuge to show that after heparin injection there was a rapid decrease in the concentrations of large lipoproteins accompanied by a concurrent increase in the concentrations of smaller, denser lipoproteins (42). Still the mechanism remained obscure. Anfinsen reviewed these data and "smelled" an enzyme. He and his newly recruited young colleagues, R.K. Brown and E. Boyle, quickly showed that heparin was indeed releasing an enzyme into the bloodstream. "Clearing factor" was a lipase (62;63). A young postdoctoral fellow, E.D. Korn, joined the group shortly thereafter and succeeded in purifying the enzyme, and he christened it "lipoprotein lipase" (64). Shannon was happy, the congressmen were happy, and Anfinsen had made his contribution to the disease-oriented mission of the National *Heart* Institute.

The lipoprotein group put together by Anfinsen expanded when the huge new research building, the Clinical Center, opened in 1954. The group in those early years included, among others, Richard J. Havel, Donald S. Fredrickson, Robert S. Gordon, Daniel Steinberg, Joseph L. Bragdon, James Baxter, Howard Goodman,

Reprinted from The Journal of Clinical Investigation, Vol. XXXIV, No. 9, pp. 1345–1353, September, 1955

THE DISTRIBUTION AND CHEMICAL COMPOSITION OF
ULTRACENTRIFUGALLY SEPARATED LIPOPROTEINS
IN HUMAN SERUM

By RICHARD J. HAVEL, HOWARD A. EDER, and JOSEPH H. BRAGDON

(*From the Laboratory of Metabolism, National Heart Institute, National Institutes of Health,
Department of Health, Education, and Welfare, Bethesda, Md.*)

(Submitted for publication January 17, 1955; accepted April 13, 1955)

FIGURE 4.8   The classic 1955 paper by Havel, Eder, and Bragdon (44) describing a relatively simple method for separating and purifying lipoprotein fractions on a preparative scale for analysis and use in metabolic studies.

and Howard A. Eder. In 1955, Havel, Eder, and Bragdon published their method for fractionating and purifying plasma lipoproteins using the preparative ultracentrifuge (44). That classic paper became to the lipoprotein field what the Lowry protein method was to biochemistry generally – the most frequently cited reference (Figure 4.8).

Over the next two decades, the work of this group and their colleagues in other Institutes, including notably Donald Frye and Robert W. Mahley, put the NIH on the map as one of the world's outstanding centers for lipoprotein and atherosclerosis research. It has continued to enjoy that reputation to this day. Anfinsen himself, having launched the enterprise, returned full time to ribonuclease but he had left his mark on the lipoprotein field – all because Shannon had gently suggested one day that "Gofman's stuff might be worth looking into."

BRINGING THE LIPOPROTEIN PACKAGE
CONCEPT INTO CLINICAL PRACTICE

One of the young tigers recruited in 1953 by Anfinsen was Donald S. Fredrickson, fresh out of residency and fellowship at Massachusetts General Hospital. Fredrickson's talent was immediately evident and he quickly rose to be the head of his own section within Anfinsen's Laboratory of Cellular Physiology and Metabolism. In 1966, he became Chief of the Molecular Diseases Branch of the National Heart Institute. Later he would go on to become Director of the National Heart, Lung and Blood Institute, then Director of the entire NIH and, finally, in 1984, President of the Howard Hughes Medical Institute, the single largest private philanthropy supporting biomedical research (Figure 4.9).

Fredrickson was persuaded of the correctness of Gofman's view that *patterns* of lipoproteins might contain valuable information beyond that given by

FIGURE 4.9   Donald S. Fredrickson, whose simple method for phenotyping patterns made the arcane world of lipoproteins more accessible to clinicians and brought more clinical investigators into the field. He went on to become Director of the National Institutes of Health and, later, President of the Howard Hughes Medical Institute. Source: courtesy of the National Institutes of Health.

measurement of the component lipids only (cholesterol and triglycerides). But he also realized that Gofman's analytical ultracentrifuge method was just too complicated and too expensive ever to be a practical clinical tool. Preparative ultracentrifugation, introduced by Havel, Eder, and Bragdon (44) in Anfinsen's laboratory, was a powerful research tool but again not practical in clinical medicine. So, when a young man named Robert S. Lees came into Fredrickson's laboratory as a postdoctoral fellow and showed him a wonderfully simple new method for separating plasma lipoproteins by paper electrophoresis (65), Fredrickson immediately saw its enormous clinical potential. Over the next few years Fredrickson and his collaborators, R.S. Lees and R.I. Levy, studied and classified the lipoprotein patterns in hundreds of patients referred to the Clinical Center at NIH (45;66;67). They found that most of them could be assigned to one of five Types. These Types provided a nomenclature and a context within which different lipoprotein disorders could be classified. Later studies would break down some of these patterns into subclasses with different underlying causes, genetic and environmental, but the availability of a relatively cheap and simple way of looking at lipoproteins sparked a wave of enthusiasm among clinicians around the world. The World Health Organization eventually adopted the Fredrickson system of classification as the international standard. It is fair to say that Fredrickson brought lipoproteins into the vocabulary of the practitioner and this undoubtedly had a major impact on the clinical management of lipoprotein abnormalities. Fredrickson's demystification of the lipoproteins helped practitioners think more clearly about the lipid hypothesis and the feasibility of preventing atherosclerosis.

MOVING FROM PHENOTYPE TO GENOTYPE

The Fredrickson classification was strictly phenotypic. Little or nothing was known about the origin and metabolism of either the individual lipoproteins or their relationship to one another. The mechanisms underlying the five Fredrickson Types remained unknown. Nor was it clear which were genetically determined and to what extent.

In 1973, Arno G. Motulsky and his colleagues at the University of Washington published the results of a heroic study involving 2,500 relatives of 149 probands who had had a myocardial infarction and who had either hypercholesterolemia or hypertriglyceridemia or both. A major driving force and first author on two of these papers was a young postdoctoral fellow, Joseph L. Goldstein. Out of this work came the first genetics-based classification of the hyperlipidemias (68–70). Three monogenic disorders were defined and these were inherited in a Mendelian dominant fashion. In many of the families, hypercholesterolemia was genetically determined but involved multiple genes. Finally there were many cases of triglyceride elevation that appeared not to be genetic. The concordance between this classification and the phenotypic classification of Fredrickson was poor, sometimes two or more phenotype patterns appearing in a single genetic disorder. Over the next years, more sophisticated studies of lipoprotein metabolism and improved methods of genetic analysis would confirm the basic correctness of this gene-based classification. The high frequency of lipoprotein abnormalities in first-degree relatives of the heart attack victims in this study again strongly suggested a causal relationship but it by no means proved it.

Nevertheless, skepticism about the lipid hypothesis continued to be the order of the day.

REFERENCES

1. Bloch, K. 1965. The biological synthesis of cholesterol. *Science* 150:19–28.
2. Bucher, N.L., Overath, P., and Lynen, F. 1960. Beta-Hydroxy-beta-methyl-glutaryl coenzyme A reductase, cleavage and condensing enzymes in relation to cholesterol formation in rat liver. *Biochim Biophys Acta* 40:491–501.
3. Lynen, F. 1966. [The biochemical basis of the biosynthesis of cholesterol and fatty acids]. *Wien Klin Wochenschr* 78:489–497.
4. Cornforth, J.W. and Popjak, G. 1958. Biosynthesis of cholesterol. *Br Med Bull* 14:221–226.

5. Popjak, G. and Cornforth, J.W. 1960. The biosynthesis of cholesterol. *Adv Enzyme Regal* 22:281–335.
6. Oncley, L.J., Scatchard, G., and Brown, H.V. 1947. Physical-chemical characteristics of certain of the proteins of normal human plasma. *J Phys and Colloid Chem* 51:184–198.
7. Barr. D.P., Russ, E.M., and Eder, H.A. 1951. Protein–lipid relationships in human plasma. II. In atherosclerosis and related conditions. *Am J Med* 11:480–488.
8. Schoenheimer, R. 1964. *The Dynamic State of Body Constituents*. Hafner Publishing Co., New York.
9. Kennedy, E.P. 2001. Hitler's gift and the era of biosynthesis. *J Biol Chem* 276:42619–42631.
10. Bloch, K. 1987. Summing up. *Ann Rev Biochem* 56:1–19.
11. Rittenberg, D. and Schoenheimer, R. 1937. Deuterium as an indicator in the study of intermediary metabolism. XI. Further studies on the biological uptake of deuterium into organic substances, with special reference to fat and cholesterol formation. *J Biol Chem* 121:235–253.
12. Sonderhoff, R. and Thomas, H. 1937. Die enzymatishe Dehydrierung der Trideutero-essigsaure. *Ann Chem* 530:195–213.
13. Bloch, K. and Rittenberg, D. 1942. On the utilization of acetic acid for cholesterol formation. *J Biol Chem* 145:625–636.
14. Little, H.N. and Bloch. K. 1950. Studies on the utilization of acetic acid for the biological synthesis of cholesterol. *J Biol Chem* 183:33–46.
15. Ottke, R.C., Tatum, E.L., Zabin, I., and Bloch, K. 1951. Isotopic acetate and isovalerate in the synthesis of ergosterol by Neirospora. *J Biol Chem* 189:429– 433.
16. Bonner, J.A.B. and Arreguin, B. 1949. The biochemistry of rubber formation in the guayule. I. Rubber formation in seedlings. *Arch Biochem* 21:109–124.
17. Robinson, J. 1934. Structure of cholesterol. *J Soc Chem Ind* 53:1062–1063.
18. Langdon, R.G. and Bloch, K. 1953. The biosynthesis of squalene. *J Biol Chem* 200:129–134.
19. Langdon, R.G. and Bloch, K. 1953. The utilization of squalene in the biosynthesis of cholesterol. *J Biol Chem* 200:135–144.
20. Tavormina, P.A., Gibbs, M.H., and Huff, J.W. 1956. The utilization of beta-hydroxy-beta-methyl-delta-valerolactone in cholesterol biosynthesis. *J Am Chem Soc* 78:4498–4499.
21. Rudney, H. 1957. The biosynthesis of beta-hydroxy-beta-methylglutaric acid. *J Biol Chem* 227:363–377.
22. Popjak, G. 1958. Biosynthesis of cholesterol and related substances. *Annu Rev Biochem* 27: 533–560.
23. Popjak, G. 1977. "As I remember it": research on biosynthesis of fatty acids, triglycerides, squalene, and cholesterol. *J Am Oil Chem Soc* 54:647A–655A.
24. Voser, W., Mijovic, M.V., Heusser, H., Jeger, O., and Ruzicka, L. 1952. Uber die Konstitution des Lanostadienols (Lanosterins) und seine Zugehorigkeit zu den Steroiden. *Helv Chim Acta* 35:2414–2430.
25. Woodward, R.B. and Bloch, K. 1953. The cyclization of squalene in cholesterol biosynthesis. *J Am Chem Soc* 75:2023–2024.
26. 1999. Professor George Joseph Popjak, MD, DSc, FRS: May 5, 1914 to December 30, 1998. *Arertioscler Thromb Vasc Biol* 19:830–831.
27. Steinberg, D. 1962. Chemotherapeutic approaches to the problem of hyperlipidemia. *Adv Pharmacol* 1:59–159.
28. Gould, R.G. and Taylor, C.B. 1950. Effect of dietary cholesterol on hepatic cholesterol biosynthesis. *Fed Proc* 9:179.
29. Gould, R.G., Taylor, C.B., Hagerman, J.S., Warner, I., and Campbell, D.J. 1953. Cholesterol metabolism. I. Effect of dietary cholesterol on the synthesis of cholesterol in dog tissue in vitro. *J Biol Chem* 201:519–528.
30. Gould, R.G. and Popjak, G. 1957. Biosynthesis of cholesterol in vivo and in vitro from DL-beta-hydroxy-beta-methyl-delta [2-14C]-valerolactone. *Biochem J* 66:51P.

31. Endo, A., Kuroda, M., and Tsujita, Y. 1976. ML-236A, ML-236B, and ML-236C, new inhibitors of cholesterogenesis produced by *Penicillium citrinium*. *J Antibiot (Tokyo)* 29:1346–1348.
32. Macheboeuf, M.A. 1929. Recherches sur les phosphoaminolipides et les sterides du serum et du plasma sanguins. *Bull Soc Chim Biol* 11:268–293.
33. Cohn, E.J., Strong, L.E., Hughes, W.L., Jr., Mulford, D.J., Ashworth, J.N., Melin, M., and Taylor, H.L. 1946. Preparation and properties of serum and plasma lipoproteins. IV. A system for the separation into fractions of the protein and lipoprotein components of biological tissues and fluids. *J Am Chem Soc* 68:459–475.
34. Russ, E.M., Eder, H.A., and Barr, D.P. 1951. Protein-lipid relationships in human plasma. I. In normal individuals. *Am J Med* 11:468–479.
35. Dobbin, E.V., Gofman, H.F., Jones, H.C., Lyon, L., and Young, C. 1951. *The Low-Fat, Low-Cholesterol Diet*. Doubleday and Company, Garden City, New York.
36. Pedersen, K.O. 1947. On a low-density lipoprotein appearing in normal human plasma. *J Phys and Colloid Chem* 51:156–163.
37. Blix, G. 1941. Electrophoresis of lipid-free blood serum. *J Biol Chem* 137:495–501.
38. Petermann, M.L. 1946. The effect of lecithinase on human serum globulins. *J Biol Chem* 162:37–42.
39. Gofman, J.W., Lindgren, F.T., and Elliott, H. 1949. Ultracentrifugal studies of lipoproteins of human serum. *J Biol Chem* 179:973–979.
40. Gofman, J.W., Lindgren, F., Elliott, H., Mantz, W., Hewitt, J., and Herring, V. 1950. The role of lipids and lipoproteins in atherosclerosis. *Science* 111:166–171.
41. Gofman, J.W. 1956. Serum lipoproteins and the evaluation of atherosclerosis. *Ann N Y Acad Sci* 64:590–595.
42. Graham, D.M., Lyon, T.P., Gofman, J.W., Jones, H.B., Yankley, A., Simonton, J., and White, S. 1951. Blood lipids and human atherosclerosis. II. The influence of heparin on lipoprotein metabolism. *Circulation* 4:666–673.
43. 1956. Evaluation of serum lipoprotein and cholesterol measurements as predictors of clinical complications of atherosclerosis; report of a cooperative study of lipoproteins and atherosclerosis. *Circulation* 14:691–742.
44. Havel, R.J., Eder, H.A., and Bragdon, J.H. 1955. The distribution and chemical composition of ultracentrifugally separated lipoproteins in human serum. *J Clin Invest* 34:1345–1353.
45. Fredrickson, D.S., Levy, R.I., and Lees, R.S. 1967. Fat transport in lipoproteins – an integrated approach to mechanisms and disorders. *N Engl J Med* 276:34–42.
46. Gofman, J.W. 1996. Atherosclerotic heart disease and cancer: looking for the "smoking guns." *FASEB J* 10:661–663.
47. Fredrickson, D.S. 1993. Phenotyping. On reaching base camp (1950–1975). *Circulation* 87:III1–15.
48. Havel, R.J. 1987. Origin, metabolic fate, and metabolic function of plasma lipoproteins. In *Hypercholesterolemia and Atherosclerosis*, D. Steinberg and J.M. Olefsky, eds. Churchill Livingstone, New York:117–142.
49. Mahley, R.W., Innerarity, T.L., Rall, S.C., Jr., and Weisgraber, K.H. 1984. Plasma lipoproteins: apolipoprotein structure and function. *J Lipid Res* 25:1277–1294.
50. Brewer, H.B., Jr. 1981. Current concepts of the molecular structure and metabolism of human apolipoproteins and lipoproteins. *Klin Wochenschr* 59:1023–1035.
51. Alaupovic, P. 1991. Apolipoprotein composition as the basis for classifying plasma lipoproteins. Characterization of ApoA- and ApoB-containing lipoprotein families. *Prog Lipid Res* 30:105–138.
52. Osborne, J.C., Jr. and Brewer, H.B., Jr. 1977. The plasma lipoproteins. *Adv Protein Chem* 31: 253–337.
53. Duff, G.L. and McMillan, G.C. 1951. Pathology of atherosclerosis. *Am J Med* 11:92–108.

54. Minnich, A. and Zilversmit, D.B. 1989. Impaired triacylglycerol catabolism in hypertriglyc-eridemia of the diabetic, cholesterol-fed rabbit: a possible mechanism for protection from ath-erosclerosis. *Biochim Biophys Acta* 1002:324–332.

55. Nordestgaard, B.G. and Zilversmit, D.B. 1988. Large lipoproteins are excluded from the arte-rial wall in diabetic cholesterol-fed rabbits. *J Lipid Res* 29:1491–1500.

56. Stender, S. and Zilversmit, D.B. 1981. Transfer of plasma lipoprotein components and of plasma proteins into aortas of cholesterol-fed rabbits. Molecular size as a determinant of plasma lipoprotein influx. *Arteriosclerosis* 1:38–49.

57. Mahley, R.W., Innerarity, T.L., Brown, M.S., Ho, Y.K., and Goldstein, J.L. 1980. Cholesteryl ester synthesis in macrophages: stimulation by beta- very low density lipoproteins from cholesterol-fed animals of several species. *J Lipid Res* 21:970–980.

58. 2001. Executive Summary of The Third Report of The National Cholesterol Education Program (NCEP) Expert Panel on Detection, Evaluation, And Treatment of High Blood Cholesterol In Adults (Adult Treatment Panel III). *JAMA* 285:2486–2497.

59. Miller, N.E. and Miller, G.J. 1975. Letter: high-density lipoprotein and atherosclerosis. *Lancet* 1:1033.

60. Hahn, P.F. 1943. Abolishment of alimentary lipemia following injection of heparin. *Science* 98:19–20.

61. Anderson, N.G. and Fawcett, B. 1950. An antichylomicronemic substance produced by heparin injections. *Proc Soc Exp Biol Med* 74:768.

62. Brown, R.K., Noyle, E., and Anfinsen, C.B. 1953. The enzymatic transformation of lipopro-teins, *J Biol Chem* 204: 423–434.

63. Brown, R.K., Boyle, E., and Anfinsen, C.B. 1953. The enzymatic transformation of lipopro-teins. *J Biol Chem* 204:423–434.

64. Korn, E.D. 1955. Clearing factor, a heparin-activated lipoprotein lipase. I. Isolation and char-acterization of the enzyme from rat heart. *J Biol Chem* 215:1–14.

65. Lees, R.S. and Hatch, F.T. 1963. Sharper separation of lipoprotein species by paper elec-trophoresis in albumin-containing buffer. *J Lab Clin Med* 61:518–528.

66. Fredrickson, D.S., Levy, R.I., and Lees, R.S. 1967. Fat transport in lipoproteins – an integrated approach to mechanisms and disorders. *N Engl J Med* 276:94–103.

67. Fredrickson, D.S., Levy, R.I., and Lees, R.S. 1967. Fat transport in lipoproteins – an integrated approach to mechanisms and disorders. *N Engl J Med* 276:215–225.

68. Goldstein, J.L., Hazzard, W.R., Schrott, H.G., Bierman, E.L., and Motulsky, A.G. 1973. Hyperlipidemia in coronary heart disease. I. Lipid levels in 500 survivors of myocardial infarc-tion. *J Clin Invest* 52:1533–1543.

69. Goldstein, J.L., Schrott, H.G., Hazzard, W.R., Bierman, E.L., and Motulsky, A.G. 1973. Hyperlipidemia in coronary heart disease. II. Genetic analysis of lipid levels in 176 families and delineation of a new inherited disorder, combined hyperlipidemia. *J Clin Invest* 52:1544–1568.

70. Hazzard, W.R., Goldstein, J.L., Schrott, M.G., Motulsky, A.G., and Bierman, E.L. 1973. Hyperlipidemia in coronary heart disease. 3. Evaluation of lipoprotein phenotypes of 156 genetically defined survivors of myocardial infarction. *J Clin Invest* 52:1569–1577.

CHAPTER 5

In Search of a Pathogenesis

THE IMPORTANCE OF UNDERSTANDING MECHANISM IN GAINING ACCEPTANCE OF A HYPOTHESIS

The lack of a well-delineated hypothesis is not necessarily a barrier to acceptance of new directions in medical practice. The classic example is John Snow's demonstration that the 1854 cholera epidemic in London was attributable to contaminants in the water. When he removed the handle from the Broad Street pump, the number of cases in the area served by that pump promptly began to wane. Exactly what was in the water that caused the cholera would not be demonstrated for more than a quarter of a century. Still the results of Snow's intervention were so dramatic that no one questioned the cause-and-effect relationship *even in the absence of an explicit hypothesis*. However, when the causal linkage is less obvious, the absence of a plausible hypothesis can be a significant deterrent to action.

To return to the case at hand, it was difficult for several reasons for physicians to accept the idea that the concentration of blood cholesterol could be a major factor determining the chances of a myocardial infarction decades down the road. As discussed in Chapter 3, it was not appreciated that the *average* blood cholesterol level in the United States, the so-called *normal* level, was actually *abnormal*. It was accelerating atherogenesis and putting a large fraction of the so-called normal population at a high risk for coronary heart disease. Also, very little was known about the structure and metabolism of these recently discovered and still mysterious cholesterol-protein complexes – the serum lipoproteins – and almost nothing was known about how they got into the vessel wall and contributed to the development of the lesions. A degree of skepticism was understandable.

The progressive enlightenment with regard to lipoprotein structure and metabolism in the post-Gofman decades together with the development of a better understanding of the cell biology of the vessel wall was critical in the fleshing out of the lipid hypothesis and is therefore an important part of this history.

The Cholesterol Wars by D. Steinberg.
Copyright © 2007 Elsevier Inc. All rights of reproduction in any form reserved.

EARLY ATTEMPTS TO DEFINE THE
PATHOGENESIS OF ATHEROSCLEROSIS

Speculations about atherogenesis date back to the 18th century and earlier but these speculations were not supported by much, if any, experimental evidence. Virchow's *insudation theory* put forward in 1856 came the closest to the mark but, like the others, it was a reconstruction based almost entirely on snapshots, i.e. on the gross and microscopic appearance of lesions at autopsy (1). He noted that lesion formation began with a thickening of the intima, which he attributed largely to "an increased imbibition of soluble components of the blood flowing past" the endothelium. He also noted the early occurrence of an increase in the numbers of subendothelial cells, an increase in their size, and the presence of lipoid substances but had no way to deduce their origin or their significance. In the absence of a suitable animal model the pathologist could not know the temporal relationships of the human lesions at autopsy.

Over 50 years later, Anitschkow beautifully described the foam cells in the lesions of his cholesterol-fed rabbits and surmised that they represented invading white blood cells (2). He reasoned this way: "one can see quite distinctly all conceivable transitional forms between the small lymphocytic and monocytic cells, on the one hand, and the large lipoid cells on the other". He also reported "lymphoid or monocytic cells frequently seem to invade the aortic wall directly from the lumen." The development of the rabbit lesions and their severity depended on how high the blood cholesterol level was and for how long it had been maintained at that level. Rabbits on normal chow showed no lesions at all so lesion initiation could be taken to begin at the time cholesterol feeding began. Anitschkow could follow the evolution of lesions as a function of time by sacrificing animals after different periods on the cholesterol-rich diet. He reported that the very earliest change, a microscopic change anteceding the advent of grossly visible lesions, was the appearance of lipid in the space between the endothelium and the underlying inner elastic membrane. (In the rabbit there are normally no cells between the endothelium and the inner elastic membrane.) In the next stage he described the appearance in that space of "cells of a polyblastic or monocytic character" that contained lipid substances "in the form of little globules." These globules were anisotropic under polarized light, exhibiting "cruciform figures typical of cholesterin [cholesterol] esters." Anitschkow, for a number of reasons, favored the view that the foam cells represented mononuclear blood cells, as mentioned above, but he had no proof, and the origin of the foam cell remained an issue for some time.

A critically important question was whether early fatty streak lesions, which are clinically benign, were the precursors of the later lesions, which are the only ones associated with clinical expression. Zinserling, in 1925, concluded from his extensive studies of autopsy material that the localization of fatty streaks was very similar to that of more advanced lesions and that the former were indeed the precursors of the advanced lesions (3). He reported finding fatty streaks frequently in babies at 6 months of age and practically always after 4 years of age. He documented that the lesions became progressively more numerous and more severe with advancing age. Many decades later the systematic study of autopsies of children dying of trauma in the USA (Pathobiologic Determinants of Atherosclerosis in Youth) came to similar conclusions (4). The importance of this relationship cannot be overemphasized. The later evolution of lesions is complex and probably affected by many more variables than the development of the fatty streak. The factors that contribute to rupture of the vulnerable plaque and the consequent terminal thrombosis need not be congruent with those that initiate lesion formation. If we want to identify the factors that initiate lesion formation, hoping to find ways to intervene and nip the disease in the bud, we should focus on the fatty streak. Readers interested in more details of the early history are referred to three chapters in the 1933 collection of essays edited by E.V. Cowdry (2;5;6).

Timothy Leary (father of the LSD guru), while a pathologist serving as Medical Examiner in Boston, made a detailed comparison of lesions in the cholesterol-fed rabbit and those in human coronary arteries and concluded, in 1934, that they were similar in most respects (7). His basic observations were in essential agreement with those of Anitschkow. However, he was struck by the large numbers of fat-filled cells in the liver of the cholesterol-fed rabbit and he also observed a number of lipid-loaded macrophages in the general circulation. From these observations he concluded that Kupffer cells loaded with lipids ("lipophages") exited the liver, entered the circulation, squeezed through the capillaries in the lungs, and penetrated the arterial wall, carrying their load of lipid in with them (8). In retrospect the circulating "lipophages" probably represented foam cells escaping from fatty streak lesions that had lost their endothelial cell cover (9). While he was wrong about the cellular mechanisms involved, his work confirmed Anitschkow's and, most important, suggested that the human and rabbit lesions were structurally similar and that lipids played a key role in both. His work was controversial at the time but helped revive interest in the possible etiologic role of blood cholesterol in the human disease.

In their influential 1951 review (10), G. Lyman Duff, the Canadian doyen of pathology, and his young collaborator, Gardner C. McMillan, agreed that the accumulation of cholesterol and other lipids was "one of the most striking morphologic features of both human atherosclerosis and experimental cholesterol atherosclerosis." They also acknowledged the potential importance of the then very new findings from Gofman's laboratory (11) correlating elevations of certain classes of lipoproteins with premature coronary heart disease. However, they concluded, as did almost all investigators at the time, that "in the vast majority of cases with or without clinically demonstrable atherosclerosis the blood cholesterol level is normal." Here was another example of how the tendency to equate *average* with *normal* led people astray. The proposition that a significant fraction of apparently disease-free people could actually be heading to a myocardial infarction because of hypercholesterolemia seemed implausible.

In 1958, Poole and Florey (12), using light microscopy, noted the presence of macrophages laden with lipid in the vessels of cholesterol-fed rabbits, both on the lumenal surface and under the endothelium. They saw some macrophages apparently penetrating between endothelial cells but, as they pointed out, there were no arrows to indicate which way they were going. Their observations supported the identification of monocyte/macrophages as the progenitors of foam cells but left open the question of just how they became loaded with lipids.

By 1963, there was a better appreciation of the complexities of lesion structure and experimental studies began on the sources of the lipid accumulating in the lesions. However, as is apparent from the presentations at a 1963 symposium on the Evolution of the Atherosclerotic Plaque (13), the role of lipoproteins remained unclear. For example, the question of whether the lipids accumulating in the lesion were the result of biosynthesis in the vascular wall or of deposition from plasma lipoproteins was still an issue to be settled. There was still disagreement about the relative contributions of smooth muscle cells and of fibroblasts to the bulk of the growing lesion because there were no unambiguous markers for the several cell types. Using the electron microscope Haust was able to show clearly that smooth muscle cells were the predominant cell type in larger, stenotic lesions, an important contribution. She also noted that some of these smooth muscle cells contained lipid droplets (14). Indeed smooth muscle cells do take up lipoproteins, remnant particles being taken up more avidly than VLDL, as later shown by Bierman *et al.* (15;16). Under the right conditions, smooth muscle cells can even express scavenger receptors (17;18) and take up modified forms of LDL. Still, only a relatively small fraction of the foam cells in the early fatty streak lesion are derived from smooth muscle cells. Haust, probably because she was focusing on the larger, space-occupying lesions, was less impressed by the lipid-loaded foam cells of macrophage origin. Another reason for the tendency to focus on smooth muscle cells and matrix deposition was that the degree of stenosis in the coronary arteries was at the time, and until quite recently, believed to be the best measure of the risk of myocardial infarction. Now it is recognized that thrombosis at the site of a ruptured plaque precipitates most infarctions. Much of the time the site of the plaque rupture is a lesion with less than 50 percent stenosis, not the tightly stenosed lesions expanding into the lumen. But, in the 1960s, the focus was on the degree of stenosis, and thus on the cellular growth and matrix deposition that caused the lesion to increase in size.

Important additional evidence favoring the lipid hypothesis was the demonstration in rabbits and in nonhuman primates that lesions could regress when the hypercholesterolemic diet was discontinued. Not only the lipid content but also, to some degree, the content of collagen and elastin could be reduced (19;20). These findings indicated that the accumulated lipid was in some way contributing to the build-up of connective tissue matrix, possibly by stimulating smooth muscle cell growth.

Later studies by Gerrity (21), using electron microscopy, and by Fowler *et al.* (22), using lipid-laden cells isolated from the lesions of cholesterol-fed rabbits, left no doubt that many or most of the lipid-laden foam cells in early lesions were derived from circulating monocytes. However, some smooth cells also imbibe lipids and have at least some of the properties of foam cells. As so often happens in science, the answer to the smooth muscle cell versus monocyte/ macrophage controversy was not either/or but both.

THE RESPONSE-TO-INJURY HYPOTHESIS AND THE MONOCLONAL HYPOTHESIS

In the early 1970s, Russell Ross, John Glomset, Laurence Harker, and colleagues at the University of Washington were trying to understand what caused smooth muscle cells at sites of atherosclerosis to proliferate so vigorously. In 1974, they made a pivotal discovery: serum from blood that had been allowed to clot ("blood serum") contained a growth factor for smooth muscle cells that was absent in serum separated from whole blood not allowed to clot ("plasma serum") (23). Similar findings had been reported for cultured fibroblasts (24;25). The factor was evidently present in blood platelets and was released when the platelets aggregated. Ross and Glomset christened the growth-promoting material "platelet-derived growth factor" (PDGF). They put these observations together with the existing evidence that mechanical injury to the endothelium could lead to platelet aggregation and intimal thickening and proposed their response-to-injury hypothesis (26;27), see Figure 5.1.

The initiating event was presumed to be some still unidentified form of "insult" to the arterial endothelium, followed by denudation with exposure of underlying matrix to which platelets adhered, releasing PDGF and possibly other growth factors. These now had access to the cells in the subendothelial space and could stimulate smooth muscle cell proliferation. Hypercholesterolemia was recognized as one possible source of injury but it was not considered to be an initiating factor in any other context. If blood lipid levels were high, there would be some accumulation of lipids in the developing lesion to be sure. However, the entry of lipids was not considered to play an obligatory role in initiation, although it might accelerate progression.

At almost precisely the same time, Ross and Harker were asking why patients with homocystinemia, due to any one of several metabolic errors, experienced frequent thrombotic episodes. Was it a direct result of the high blood homocystine

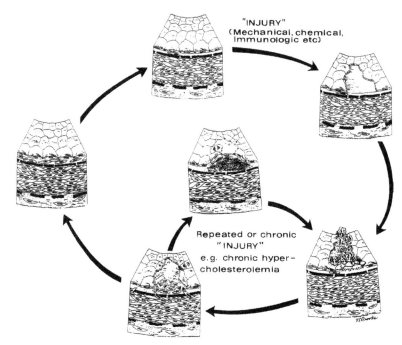

FIGURE 5.1   Schema of the Ross/Glomset response-to-injury hypothesis. Source: reprinted with permission from reference (26). Copyright 1976, Massachusetts Medical Society. All rights reserved.

levels or was it due to some other end result of the metabolic error? They showed that infusing homocystine intravenously *in vivo* into baboons caused extensive endothelial denudation, a marked increase in endothelial cell proliferation and a threefold increase in platelet turnover (28;29). They concluded, "homocystine-induced endothelial injury resulted in arteriosclerosis through platelet-mediated intimal proliferation of smooth muscle cells." These findings in an animal model that mimicked a human disease associated with vascular complications, taken together with the newly discovered platelet-derived growth factor for smooth muscle cells, made a compelling story.

The impact of the response-to-injury hypothesis was great. Here was a way to account for smooth muscle proliferation, a prominent feature of the growing plaque. The two-part review published by Ross and Glomset in the *New England Journal of Medicine* in 1976 (26;27), quickly became the standard reference on the pathogenesis of atherosclerosis. Indeed, Ross and Glomset's work proved to be importantly heuristic. It was the first to show how new approaches from cell biology could be brought to bear on the problem of atherogenesis.

As first proposed, the theory was that some "injury" caused endothelial desquamation, allowing platelets to adhere to the exposed intimal collagen and to release PDGF (see Figure 5.1). Repetition of such injuries over the years led to the complex later lesions. Later studies, however, showed that the endothelial layer overlying the initial fatty streak lesion was actually unbroken (9;12;21;30). These and other findings required later revisions of Ross's theory, substituting functional injury for structural injury. However, the focus on the penetration of leukocytes and their interactions with the cells of the artery wall stimulated a good deal of interest in atherogenesis. In his last update of the hypothesis, published just a few months before his untimely death at age 69, Ross reviewed the factors that might be responsible for the "injury" to the endothelium (31). While acknowledging the possible role of hyperlipoproteinemia as an initiating factor that might promote inflammation in the artery, he gave it no greater weight than homocystinemia, hypertension, infection, or other potential pro-inflammatory factors.

At about the same time that Ross and Glomset were developing their hypothesis, Benditt and Benditt, in the same department at the University of Washington, put forward their monoclonal hypothesis of atherogenesis (32). What they proposed was that the smooth muscle cells accumulating in any given localized atherosclerotic lesion had their origin from a single cell that had somehow been triggered to grow rapidly enough to become a benign tumor. In other words the atheroma was somewhat analogous to the leiomyoma, which had already been shown to exhibit monoclonality (33).

So we had the paradoxical situation that during the same years that the evidence for hypercholesterolemia as a primary causative factor was accumulating, the two most widely accepted hypotheses for the pathogenesis of atherosclerosis focused on smooth muscle cell proliferation and barely mentioned lipoproteins. Lipid accumulation was recognized to occur, of course, but it was regarded almost as an epiphenomenon. The response-to-injury hypothesis and the monoclonal hypothesis provided attractive schemes of pathogenesis and neither appeared to involve hypercholesterolemia in any substantive way. Little wonder then that the skeptics could easily discount the lipid hypothesis as "case not proved." Nevertheless, both sides could agree on a basic sequence of events that initiated the fatty streak lesion, as shown in Figure 5.2.

UNDERSTANDING THE ROLE OF HDL AS AN ATHEROPROTECTIVE FACTOR

The atheroprotection afforded by HDL was evident long before it was understood mechanistically. The conventional view of the lipoprotein/lipid transport system, until the 1960s, was that the liver secreted VLDL; the VLDL was acted on by lipoprotein lipase to be degraded stepwise to IDL and LDL. During this processing, the bulk of the VLDL triglycerides and phospholipids were delivered

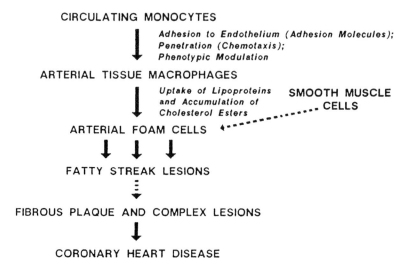

FIGURE 5.2   A sequence of events for the initiation and progression of atherosclerotic lesions generally agreed upon by most investigators as of 1980.

to the peripheral tissues. It was considered that all, or almost all, of the LDL cholesterol must be returned to the liver from whence it came for disposal in the bile. In fact this seemed to be a logical necessity because peripheral tissues (excepting endocrine tissues) cannot degrade cholesterol, yet all of them do synthesize it at some rate. When later the sites of irreversible degradation of protein-labeled LDL were directly measured it was found that as much as 30 percent of that degradation occurred in nonendocrine extrahepatic tissues (34), underscoring further the necessity for some mechanism of returning cholesterol from peripheral tissues back to the liver, i.e. "reverse cholesterol transport." Of greatest importance in the present context, arteries take up LDL at a high rate (expressed per unit wet weight) (35) so the relevance of a reverse cholesterol transport pathway for controlling cholesterol accumulation in lesions and, potentially, facilitating regression of atherosclerosis became apparent.

   John A. Glomset was the first to propose an explicit mechanism for reverse cholesterol transport, a proposal growing out of his studies of lecithin-cholesterol acyl transferase (36;37). Sperry had discovered as early as 1935 that when plasma or serum was simply incubated overnight at 37°C, the amount of cholesterol ester in it increased considerably and a corresponding amount of free cholesterol disappeared (38). Preheating the plasma at 55–60°C prevented all these changes, so Sperry correctly concluded that a plasma enzyme was involved and initially proposed that it was a cholesterol esterase. Later studies in several laboratories showed that lecithin was the source of the fatty acids appearing in the newly synthesized cholesterol esters (reviewed by Glomset (37)).

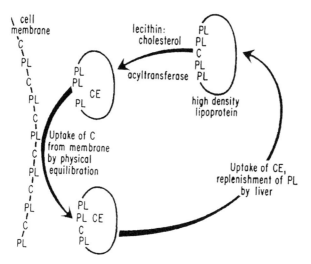

FIGURE 5.3 Schema proposed by Glomset by which free cholesterol could be mobilized from tissues and transported to the liver for excretion in bile. The cycle depended on the action of lecithin-cholesterol acyl transferase. Source: reproduced from reference (37) with permission.

The exact molecular mechanism still remained unclear. It remained for Glomset and co-workers to show that a single enzyme – lecithin-cholesterol acyltransferase (LCAT) – accounted for the Sperry observations (39;40). Basically what they proposed (Figure 5.3) was that LCAT could convert free cholesterol on the surface of a lipoprotein to its ester form; the ester form, being more hydrophobic, would transfer into the hydrophobic core of the lipoprotein; this would create a "vacancy" on the lipoprotein surface to which free cholesterol molecules from the tissue could then attach. Repeated cycles of this kind could result in a net removal of cholesterol from the tissues. The lipoproteins would then be taken up by the liver with excretion of the cholesterol in the bile.

In his review, Glomset called attention to a clever experiment done independently by Murphy in 1962 that supported his proposed mechanism. Murphy showed that when red blood cells were incubated with serum that had previously been incubated overnight at 37°C, thus allowing a large fraction of the lipoprotein free cholesterol to be converted to ester, there was a net transfer of free cholesterol from the red cells into the serum lipoproteins; if the serum had not been preincubated, no such transfer occurred (41). These were the basic findings that led ultimately to an elucidation of the complex system of reverse cholesterol transport, a major mechanism by which HDL exerts it anti-atherogenic action.

While there was general agreement that Glomset's hypothesis was essentially correct, direct evidence for the operation of reverse cholesterol transport *in vivo* was a long time coming. In his Chairman's remarks at a 1977 Symposium on

High Density Lipoproteins and Atherosclerosis (42), Robert I. Levy summed it up this way: "All agreed that in [terms of] understanding HDL, its role and its determinants, we were at best at 'the end of the beginning.' HDL is clearly a lipoprotein species of importance. Now all we have to do is find out why and how to control it."

In part, the difficulty stemmed from the rapid exchanges of both apoproteins and lipids, between HDL and other lipoprotein classes, and between the several species of HDL. In the rat, however, there is very little apoA-I outside the HDL fraction and very little cholesterol ester exchange activity (43). Using HDL with the apoA-I labeled with ^{125}I in a nondegradeable marker and simultaneously labeled with ^{3}H-cholesteryl ether (not a substrate for cholesterol esterases) it was shown that HDL is taken up into the liver by two different pathways;

1. as intact HDL particles (i.e. both labels entering at the same rate); and
2. by a selective mechanism allowing the cholesterol ester to enter independently of uptake of the holo-HDL particles (i.e.^{3}H entering more rapidly than^{125}I).

At the time, the mechanism of this "selective cholesterol ester uptake" was not known. Then, in 1996, Acton *et al.* in the laboratory of Monty Krieger identified scavenger receptor B-1 as an HDL-binding receptor that could engineer the selective uptake of cholesterol esters from HDL without uptake of the apoprotein (44). We currently have a much clearer idea of the mechanisms by which cholesterol in peripheral cells, especially macrophage foam cells, is transferred to HDL, involving SR-B1 and at least two specific ATP-binding cassette transporters – ABCA1 and ABCG1 (see review by Barter and Rye (45)).

AN UPDATE ON HDL AS A TARGET FOR INTERVENTION

The negative correlation between HDL cholesterol levels and risk is as strong as or even stronger than the positive correlation between LDL cholesterol levels and risk. To the extent that the protective effect of HDL is due to its ability to enhance removal of cholesterol from arterial lesions, the data added to the strength of the basic lipid hypothesis. To grossly oversimplify: LDL delivers cholesterol to the developing lesion at some rate and HDL removes it at some rate. If the rates are equal there should be no net cholesterol deposition; if the LDL level is very low, then even a relatively low HDL level should be enough to avoid net deposition; in fact, if the LDL level were zero or close to it, there would be no net deposition no matter how low the HDL. The latter seems to be the case in the mouse. Gene targeting to knock out apoA-I does not evoke any arterial lesions in a wild-type, atherosclerosis-resistant strain of mouse unless at the same time the level of LDL of the mouse is elevated somehow (e.g. by introducing the gene for human apoB) (46). Much or most of the protective effect of HDL seems to be attributable to its role in reverse cholesterol transport but there is increasing evidence now that more is involved, namely, an anti-inflammatory component (45;47). This would account for situations in which the protective effect of HDL does not parallel the plasma concentration of HDL.

A very exciting recent development is the discovery that some of the anti-inflammatory effects of HDL can be mimicked by small peptides derived from the sequence of apoA-I and that these can be effective even when administered orally if they are synthesized from D- rather than L-amino acids (48;49).

Even before these mechanisms of action had been elucidated, HDL levels were being used as a guide to the risk level in individual patients, and therefore as a guide to how aggressively they should be treated. Knowing now that the high HDL was not just a secondary marker for some other anti-atherogenic factor made it a legitimate target for direct intervention. Because nicotinic acid, more than any of the other drugs in common use, raises HDL (by as much as 30 percent), these new findings renewed interest in using it clinically and the pharmaceutical industry set out to find, and did, more acceptable formulations. They also began to explore more actively the possibility of finding drugs that would raise HDL levels. One of these, torcetrapib, an inhibitor of cholesterol ester-phospholipid transferase (CEPT), made it all the way to Phase III clinical trials (50;51). In principle, a combination treatment with a statin plus an HDL-raising drug should be ideal. However, it remains to be seen whether and to what extent the increase in HDL levels induced by torcetrapib or other HDL-raising drugs is mirrored by an effect on coronary heart disease event rates.[1]

DISCOVERY OF THE LDL RECEPTOR: THE REMARKABLE PARTNERSHIP OF BROWN AND GOLDSTEIN

Without question the discovery of the nature of the defective gene in familial hypercholesterolemia (FH) by Goldstein and Brown was a major milestone in the lipoprotein field (52–54). In a remarkable series of elegant and insightful papers published in the 1970s and 1980s, they established that the cellular uptake of LDL absolutely requires the LDL receptor. In the complete absence of a functional receptor, the LDL cholesterol concentration can build up to 800–1,000 mg/dl. Since FH was clearly a monogenic disorder it could now be said that the high LDL levels, secondary to the lack of LDL receptor function, must be the immediate and *sufficient* cause of atherosclerosis in these patients, including the devastating myocardial infarctions that sometimes occur as early as the first decade of life. The importance of the Goldstein/Brown work in supporting the lipid hypothesis cannot be overstated. They were awarded the Nobel Prize in Physiology or Medicine in 1985. How they became lifetime collaborators and how they made this seminal discovery makes a fascinating story.

They first met and learned to appreciate one another when they both served their internship and residency in medicine at Massachusetts General Hospital, from 1966 to 1968. Then they both spent the next 2 years at the National Institutes of Health, Goldstein working with Marshall W. Nirenberg, winner of a Nobel Prize in Physiology or Medicine in 1968, and Brown with Earl R. Stadtman, probably the most brilliant enzymologist at NIH (or anywhere else for that matter). Working as they were in different laboratories, there was no opportunity to do collaborative biochemical research at that time. However, they both wanted ultimately to do research on metabolic diseases and both were intrigued by the still mysterious disorder of familial hypercholesterolemia. Goldstein, as a Clinical Associate responsible for the medical care of Dr. Donald S. Fredrickson's research patients in the Clinical Center, saw these fascinating cases close up and discussed them with Brown. The two of them shared many common scientific interests but the seeds of what would become a lifelong partnership were sown not in the laboratory, but over the bridge table. Both were duplicate bridge fiends.

Goldstein did his medical training at Southwestern Medical School at the University of Texas Health Science Center in Dallas. Donald W. Seldin, Chair of the Department of Internal Medicine, had a keen eye for talent and he saw that Goldstein was simply brilliant. Seldin was following a "grow your own" strategy to build the research strengths of his department, which was at the time already the most outstanding department in the medical school and one of the best in the country. So during Goldstein's senior year Seldin "recruited" him. If Goldstein would agree to obtain graduate training in human genetics, Seldin would guide him through his residency and postdoctoral training and guarantee him a faculty position as head of a division of genetics on his return. Seldin had, and still has, the highest respect for excellence. Seldin also had, and still has, an intensity and charm that is irresistible. His was an offer that Goldstein could not refuse.

Brown got his M.D. at the University of Pennsylvania, where he was number one in his class. As a medical student he dove into research, spending three summers in the Smith-Kline laboratories. A rotation in Albert Winegrad's laboratory at Penn aroused his interest in lipid metabolism and in a research career. James Wyngaarden, Chairman of Medicine at Penn at the time, helped engineer the residency for Brown at Massachusetts General Hospital.

During their 2 years as Clinical Associates, Goldstein tried to "do a Don Seldin" on Brown, extolling the virtues of Southwestern Medical School in general and of Don Seldin in particular. "Come to Dallas and together we'll solve the mysteries of familial hypercholesterolemia" may have been the bottom line. It worked. Brown, after working another year at NIH, in Earl Stadtman's laboratory, accepted Don Seldin's offer of a faculty position in the Division of Gastroenterology. Brown, working with with Dana, Dietschy, and Siperstein, partially purified and characterized the HMGCoA reductase from liver (55), setting the stage for what was to follow when Goldstein returned to Southwestern.

Goldstein, after his 2-year postdoctoral fellowship with Arno G. Motulsky, a world-class human geneticist at the University of Washington, returned to Dallas per agreement with Seldin. The Brown and Goldstein collaboration got under way … and it is still going. They have continued to work smoothly as a two-man team now for almost 35 years.

GOLDSTEIN AND BROWN START THEIR SEARCH FOR THE FAULTY GENE IN FAMILIAL HYPERCHOLESTEROLEMIA

When they started their collaboration Brown and Goldstein knew that familial hypercholesterolemia (FH) was due to a single gene mutation; the earlier studies of Wilkinson *et al.* (56), Adlersberg *et al.* (57) and Khachadurian (58) had established that. But which gene? Marvin D. Siperstein, a senior member of the Department of Medicine in Dallas, was working at the time on the rate-limiting enzyme in the synthesis of cholesterol, HMGCoA reductase. Goldstein and Brown picked this enzyme as a good starting place. They postulated that the cells in patients with FH might be producing cholesterol at an abnormally high rate secondary to a genetic flaw in the regulation of the reductase enzyme.

CHOLESTEROL SYNTHESIS IN SKIN FIBROBLASTS IN CELL CULTURE

The technique of growing cells in culture was still relatively new in 1972. The notion that cultured cells might allow pinpointing of metabolic errors was still newer. At that time the liver was believed to be both the source of the blood lipoproteins and the site at which they were removed from the blood. Brown and Goldstein would therefore have liked to study liver cells in culture, but it was difficult to justify the risks associated with liver biopsies purely for research purposes. On the other hand, human skin fibroblast cells were readily grown in culture and if the gene defect were global it might be apparent in skin fibroblasts. Several gene abnormalities underlying metabolic disorders had already been discovered and characterized using skin fibroblasts so Goldstein and Brown decided to give it a try. Their first hypothesis was that the rate of cholesterol synthesis would be abnormally high. They would assay the activity of the reductase enzyme and take that as a measure of the rate of cholesterol synthesis.

In their very first set of experiments they found that in normal cells grown in the presence of serum the rate of cholesterol synthesis was low. However, when the serum was removed and the cells were incubated overnight in a simple, protein-free medium, the rate of cholesterol synthesis rose sharply, as much as tenfold. They showed that the suppressive activity of the serum on the normal cells resided in the LDL fraction.

In contrast, cells from patients with FH always showed a high rate of cholesterol synthesis, even in the presence of serum. Moreover, the addition of LDL to the medium, which reduced synthesis tenfold in normal cells, had absolutely no inhibitory effect in FH cells. At that point Brown and Goldstein postulated that the gene defect must be in some "hitherto unidentified gene whose product is necessary for mediation of feedback control by lipoproteins." It was known that cholesterol in the diet suppressed the rate of cholesterol synthesis and that cholesterol synthesis in tissues was suppressed by cholesterol in the incubation medium (59;60). So at this point they were presumably thinking about a system for regulating the synthesis of cholesterol that was faulty in FH patients because one of the genes involved in that "feedback" regulation was defective.

THE PROBLEM IS GETTING LDL INSIDE THE CELLS!

Then they discovered that the FH cells, while not responsive to LDL in the medium, responded nicely if pure cholesterol (dissolved in alcohol) was added to the culture medium instead of LDL. The response of the FH cells to free cholesterol in the medium was no different from that of normal cells. In other words the FH cells could respond just as well as normal cells *provided* the cholesterol got into the cells. The FH cells could not take up cholesterol when it was offered as a component of LDL, but could respond normally to cholesterol once it got inside the cell. That is how the concept of the LDL receptor was born. It provided a mechanism for the transfer of LDL, with its cholesterol, from the surrounding medium to the inside of the cell. Goldstein and Brown went on to characterize the receptor, delineate the "LDL receptor pathway" and, ultimately, to clone the receptor.

These were elegantly simple experiments, carefully conducted and correctly interpreted. This was the first transport receptor to be characterized. Hormone receptors also bind their respective ligands with high affinity but they regulate the cell's metabolism by molecular signaling ("second messenger" systems). The concept of receptor-mediated endocytosis represented a major contribution to biology in general. Large numbers of other receptors that function in this way have, of course, since been identified and characterized. The LDL receptor of Brown and Goldstein was the prototype (61). The schema shown in Figure 5.4 became arguably the most reproduced "logo" in cell biology in the 20th century.

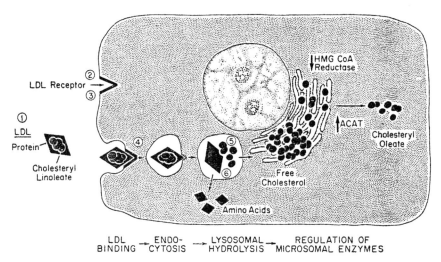

FIGURE 5.4   The LDL receptor pathway for endocytosis as originally proposed by Brown and
Goldstein in 1976. Source: reprinted with permission from reference (61); copyright 1976 AAAS.

These results considerably strengthened the lipid hypothesis. The key point
was that a monogenic defect was sufficient to raise plasma LDL markedly and
that apparently was in turn sufficient to cause atherosclerosis. The atherosclerosis was not the result of some other pathway affected by the LDL receptor gene
mutation. There was, however, one caveat. Patients with familial hypercholesterolemia have elevated levels of intermediate density lipoproteins (IDL) as
well as of LDL. This is because normally a large fraction of the IDL and (some
VLDL) is taken up, like LDL, by way of the LDL receptor. But that uptake is
linked not to binding of apoprotein B but of the apoprotein E on the lipoprotein particles (62). Also, the apoE-rich IDL (β-VLDL) are avidly taken up by
macrophages and could account for foam cell formation, as mentioned above
(63;64). So at one point the notion was seriously entertained that it might be
the IDL rather than the LDL that was mainly responsible for the atherosclerosis in FH. The later discovery of patients with defective apolipoprotein B-100,
caused by a mutation at residue 3500, resolved the question (65;66). The LDL
in these patients binds very poorly to the LDL receptor and so LDL levels can
build up to values almost as high as those seen when the LDL receptor itself is
defective. But these patients do not have any elevation of IDL levels because
their LDL receptors are normal and the affinity of their remnant lipoproteins for
the LDL receptor, which depends on their apoE content, is normal. As pointed
out by Myant (67), their cholesterol levels, while variable, can be comparable
to those in FH and the severity of their atherosclerosis is similar. In this way
the 3500 apoprotein B mutation, raising LDL levels in an entirely different way

and without causing a build-up of IDL, further supported the lipid hypothesis and pinpointed LDL as the key atherogenic lipoprotein. This is not to say that IDL never plays a role; when present at high levels, as in patients with the apoE2/E2 phenotype, IDL can certainly play an atherogenic role.

BROWN AND GOLDSTEIN: AN APPRECIATION

The Goldstein–Brown partnership published its first joint paper in 1973. Seldom has there been such a fruitful blending of two talents. Over the next 12 years they published an average of 10–12 papers a year and every one of them was highly significant. Many scientists publish (and, sad to say, even republish) minor findings to pad their bibliographies. Brown and Goldstein never did that. They were quickly recognized by everyone in the field as a major new force. Some spoke of them with amazement (and probably a touch of jealousy) as the Dallas Paper-of-the-Month Club. It was said that they sustained the breathtaking pace of their research by dividing responsibilities on a rotating basis: one would run the laboratory while the other wrote the papers! Every original paper they wrote was jointly co-authored – and they alternated the order of authorship religiously between Brown–Goldstein, and Goldstein–Brown. This was a partnership in a class with Gilbert and Sullivan, Rodgers and Hammerstein, or, more aptly, Stein and Moore (who shared the Chemistry Nobel Prize in 1972 with C.B. Anfinsen). Goldstein and Brown were made full Professors by age 36, elected to the National Academy of Sciences at age 40 and won the Nobel Prize in 1985 at age 45. It was my honor to introduce them at a conference in San Diego in 2005 and I pointed out that the Nobel Prize had not damaged their productivity, as it often has for other winners. In 2003, they won the prestigious Albany Prize and the citation specifically declared that it was being awarded for work done after 1985 (Figure 5.5).

Before leaving the "B and G" story it should be said that they have been unfailingly generous to colleagues and trainees. Almost every one of their trainees has gone on to hold a major chair. I never had the pleasure of collaborating directly with them but I have my own direct experience of their magnanimity. In 1975, Nicholas B. Myant at Hammersmith Hospital arranged a Ciba Symposium in London at which Brown and Goldstein and our group from La Jolla were invited to present. Olga and Yechezkiel Stein from Jerusalem, pioneers in the study of lipoproteins and atherosclerosis, were at the time visiting scientists in my laboratory in La Jolla. Together with my postdoctoral fellow, David B. Weinstein, we had studied LDL metabolism in cultured cells that Myant had sent us from one of his patients with homozygous FH. Our data nicely confirmed the basic Brown–Goldstein finding: these cells could not degrade LDL. However, we found only a modest defect in LDL binding and proposed that in this particular patient (and possibly in other FH patients) the defect was not in the binding to the cell surface but in the mechanisms by which the LDL, once

FIGURE 5.5 Joseph L. Goldstein and Michael S. Brown at the time they were awarded the Nobel Prize in Physiology or Medicine in 1985. Source: photo courtesy of Drs Brown and Goldstein.

bound to the surface, was internalized. At the London conference, when I got up to present our results I put the names of our four authors on the blackboard – Stein, Weinstein, Stein, and Steinberg – and referred to it as the "Vierstein" paper (68). When Brown got up to make his presentation he said that in Dallas they, unfortunately, had only one Stein – and, quick on the draw as always, scrawled "Einstein" on the blackboard.

It turned out that they were absolutely right with only one Stein, while we were wrong with four Steins. We had been misled by the general "stickiness" of LDL and were measuring a lot of nonspecific (irrelevant) binding that masked the binding defect that was actually there. Goldstein and Brown's subsequent work with the same cell line from Myant's patient showed that there was a clear deficiency in binding in addition to the defect in degradation. Ironically, Brown and Goldstein later discovered an FH patient in whom the receptor did actually bind normally but failed to internalize (69). Throughout this contretemps, Goldstein and Brown were generously nonjudgmental, sharing data with us and offering advice. Nor did it stop them from inviting me to Stockholm in 1985 as one of the six colleagues the winners are allowed to invite as special guests of the Nobel Committee. They are good at forgiving.

DISCOVERY OF THE SCAVENGER RECEPTOR ON MACROPHAGES

The discovery of the LDL receptor by Brown and Goldstein represented a turning point in the history of lipoprotein research (54) but equally important was their discovery of the scavenger receptor on the macrophage (70;71) and the nature of the ligands for it (72). They were struck by the fact that most of the cells from patients with homozygous FH take up LDL at very low rates. Indeed, the slow uptake by the liver accounts for the very high steady state concentrations of LDL in their blood. Yet the cells in xanthomas and in athero-sclerotic lesions are heavily loaded with cholesterol, suggesting that they might be taking up LDL rapidly, even though these patients have no functional LDL receptors. Therefore, the uptake into the macrophages had to be by some alternative mechanism(s).

Knowing that foam cells in lesions were largely derived from circulating monocytes, they tried to generate foam cells *in vitro* by incubating mouse peritoneal macrophages or circulating monocytes with high concentrations of LDL. Even at very high concentrations uptake was slow and no foam cells developed. Since the ultimate source of the cholesterol stored in xanthomas and in arterial lesions had to be plasma LDL, they reasoned that the circulating LDL must undergo some modification and that it was the modified form that entered the macrophages. They explored a number of chemical and enzymatic modifications of LDL but the only one that worked was chemical acetylation. Acetyl-LDL bound with high affinity to macrophages and was taken up rapidly enough to increase strikingly the cell content of cholesterol. Moreover, this uptake had all the earmarks of a receptor-mediated process and they christened the receptor the acetyl-LDL receptor. However, there was no evidence that acetyl-LDL was generated *in vivo* and they considered it unlikely that it would be. Fogelman *et al.* (73) showed that malondialdehyde-treated LDL bound with high affinity to macrophages and suggested that such a modification might occur *in vivo* when platelets aggregated. However, there was again no evidence that such a modification ever occurred *in vivo*. Other chemical modifications involving conjugation of lysine residues with malondialdehyde (73) or by acetoacetylation (74) were shown to mimic acetylation but none of these modifications of LDL was shown to occur *in vivo*. The search for the modified LDL postulated by Goldstein and Brown went on and several candidates emerged.

OXIDATIVELY MODIFIED LDL AND ATHEROGENESIS

In 1979, Henriksen *et al.* in Oslo (75) and Hessler *et al.* in Cleveland (76) independently observed that cells cultured in the presence of native LDL and in the absence of serum underwent severe damage, beginning to die within 24 hours. This cytotoxicity was inhibited by serum or by HDL. Henriksen, interested in the mechanisms underlying this cytotoxicity, came as a visiting scientist to the Specialized Center of Research on Arteriosclerosis (SCOR) in La Jolla. There

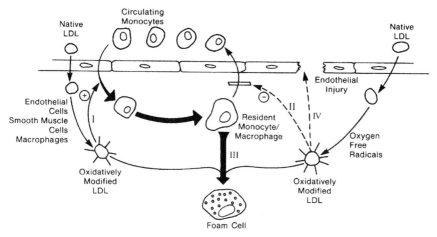

FIGURE 5.6   Schema showing how oxidized LDL might be involved as as initiator of atherosclerosis, based on the work of Henriksen *et al.* (75;77), Hessler *et al.* (76), and Quinn *et al.* (80). Source: reprinted with permission from reference 80.

he showed, in 1981, that during the incubation of LDL with cultured endothelial cells the LDL was drastically altered, becoming denser, more electronegative, and, most important, becoming a high affinity ligand for some receptor or receptors on the macrophage. This so-called "endothelial cell-modified LDL" could increase the cholesterol content of the macrophage and it was added to the list of LDL modifications that might solve the paradox of how macrophages became foam cells (77). Then Hessler and collaborators in Cleveland demonstrated that the changes induced in LDL by incubation with endothelial cells were due to free radical modification (78). Steinbrecher *et al.* in La Jolla independently came to the same conclusion with regard to the mechanism by which endothelial cells converted LDL into a ligand for macrophage receptors (79). Addition of vitamin E or a low concentration of normal serum blocked both the oxidation and the conversion of LDL to its more atherogenic form. Based on these findings, together with the later finding that oxidized LDL was chemotactic for circulating monocytes but actually inhibited the motility of activated macrophages (80), the hypothetical scheme for lesion initiation shown in Figure 5.6 was proposed.

If the hypothesis was correct, then inhibition of the oxidation of LDL might in theory slow the progression of atherosclerosis in experimental animal models and, hopefully, in humans. But which antioxidant to try? Chance stepped in to point the way.

At the same time that the oxidative modification hypothesis was taking shape, another group in the La Jolla SCOR laboratories was trying to shed light on the mechanism by which probucol, already in clinical use, reduced LDL levels (81).

LDL samples were always being exchanged among the collaborating groups and, this particular week, one of the LDL samples used for probing LDL oxidation came from a rabbit under treatment with probucol. That LDL refused to be oxidized! Two and two got put together and it was quickly shown that adding probucol directly to the medium was a fine way to block endothelial cell-induced LDL oxidation (82). Because probucol had already been approved for clinical use it was decided to go directly to it as a test of the hypothesis. Carew *et al.* (83) demonstrated that probucol slowed the progression of lesions in the LDL receptor-deficient rabbit by more than 50 percent. Most importantly, they showed that the effect was not due to its LDL-lowering effect. A second control group of rabbits was treated with just enough lovastatin to match the LDL levels reached in the probucol-treated group. The levels of probucol reached in the LDL of the experimental rabbits was sufficient to protect their LDL from oxidation *ex vivo* and the anti-atherogenic effect was assumed to reflect the antioxidative action of the drug. Independent studies by Kita and co-workers in Kyoto also showed an anti-atherogenic effect in similar experiments but without a secondary control group to assess the relative contribution of LDL lowering per se (84). The observations were attributed to the antioxidant properties of probucol, but additional mechanisms could not be ruled out. A very recent report by Wu *et al.* in Roland Stocker's laboratory suggests that the mechanism may be complex, possibly relating to activation of heme oxygenase (85).

The new findings on oxidized LDL were strongly heuristic, stimulating a burst of activity in many laboratories. Over the next 10 years over 300 papers were published relating to oxidized LDL (OxLDL) and its possible role in atherogenesis; the number now stands at over 3,000.

The oxidative modification hypothesis was strongly supported by a number of lines of evidence, both *in vitro* and *in vivo*. No attempt will be made here to review this large body of evidence. The interested reader is referred to several comprehensive reviews (86–90).

Briefly:

1. OxLDL can convert macrophages to foam cells through uptake mediated by scavenger receptors.

2. OxLDL has many other biological effects that would accelerate atherogenesis. For example, it can be cytotoxic; it increases the expression of monocyte-selective adhesion molecules on the endothelium; it increases the expression of chemoattractants; it stimulates growth of smooth muscle cells.

3. OxLDL can be demonstrated in the plasma and in the atherosclerotic lesions of animals and humans.

4. Antibodies recognizing OxLDL are present in the plasma of normal animals and humans; the titers are higher in animals and humans with severe atherosclerosis.

5. Atherosclerosis in animal models can be strikingly inhibited by administration of an appropriate antioxidant compound (e.g. probucol; probucol analogues; diphenylphenylenediamine; butylated hydroxytoluene; vitamin E). The mechanism(s) by which probucol acts is discussed in more detail in Chapter 6.

6. Lipoxygenases can convert LDL to OxLDL *in vitro*; over expression of lipoxygenase increases and lipoxygenase gene-targeting decreases atherogenesis.

CLINICAL TRIALS

The epidemiologic evidence and the evidence in experimental animal models were sufficiently persuasive to lead to the recommendation in 1992 that clinical intervention trials with antioxidants be undertaken (91). However, the trials to date, most of them using either beta-carotene or vitamin E as the antioxidant, have been negative (92;93). The possibility that other antioxidants, possibly started earlier in life or acting by other mechanisms, may prove effective has not been ruled out (90;92). Here we want only to record how the availability of an explicit hypothesis linking a modification of LDL directly to atherogenesis helped enlist support for the recognition of hypercholesterolemia as a centrally important causative factor in coronary heart disease, no matter precisely how the hypercholesterolemia led to the atherosclerotic lesion.

OTHER LIGANDS AND OTHER MACROPHAGE RECEPTORS THAT MAY CONTRIBUTE TO FOAM CELL FORMATION

Lipoprotein remnant particles rich in apoprotein E can effectively load macrophages with cholesterol and could play a role in foam cell generation (63). These lipoproteins are apparently recognized by the LDL receptor rather than the scavenger receptor so they would not account for foam cell formation in familial hypercholesterolemia patients since they have no functional LDL receptors. However, they certainly could play a role in other cases of hypercholesterolemia. Further research is needed.

Aggregated LDL is much more rapidly taken up by macrophages than native LDL but again the uptake appears to occur via the LDL receptor (94). Aggregation of LDL in the subendothelial space has been demonstrated (95) and this may be encouraged by the proteoglycans in the artery wall to which LDL binds avidly (96).

Immune complexes of LDL with antibodies can enhance macrophage uptake via the F_c receptor (97), another possible mechanism to account for foam cell formation, in this case even in the absence of LDL receptors.

Finally, it should be noted that macrophages express more than one "scavenger receptor." The first such receptor, SRA, was characterized by Kodama *et al.* in Krieger's laboratory (98). Early studies by Sparrow *et al.* (99) provided evidence that mouse macrophages must contain one or more additional receptors recognizing oxidized LDL and, in 1993, Endemann *et al.* (100) characterized and cloned CD36 as a second oxidized LDL receptor. Subsequent studies show that CD36 is the dominant receptor for uptake of oxidized LDL (101;102) while SRA is mainly responsible for the uptake of acetyl-LDL. The amelioration of atherosclerosis by knocking out of either SRA or CD36 receptors for modified forms of LDL, added strength to the case against LDL and supported the oxidative modification hypothesis (102;103). The finding that bone marrow transplantation using CD36 negative cells reduced severity of atherosclerosis was consonant with these results (103A). On the other hand, a recent study, using somewhat different methods, failed to find any effect of CD36 or SRA knockout on atherogenesis for reasons not yet clear (103B, 103C). Scavenger receptors recognize and take up oxidized LDL but do not recognize native LDL, β-VLDL or aggregated LDL, the other modified forms of LDL that have been reported to lead to foam cell formation *in vitro*. The results imply that foam cell generation reflects uptake of oxidized LDL (and possibly other similarly modified forms of LDL) and that foam cell generation is essential for atherogenesis, at least in these mouse models.

INFLAMMATION IN THE PATHOGENESIS OF ATHEROSCLEROSIS

From one point of view the role of inflammation in atherogenesis might be considered a side issue in a book devoted to tracing the evolution of the lipid hypothesis. For a time the "inflammation hypothesis" was regarded as a competing alternative to the lipid hypothesis. Proponents tagged atherosclerosis as an inflammatory disease and barely mentioned lipids (104;105). This false dichotomization may have delayed the acceptance of the lipid hypothesis. Today it is clear that the two hypotheses are by no means mutually exclusive (106). Note that studies of the inflammatory events are almost all done in mice (or other animal models) that have lesions only because they were made hypercholesterolemic, and they are kept hypercholesterolemic throughout the study. Thus, the inflammatory processes are always preceded by and accompanied by hypercholesterolemia. Hypercholesterolemia leads rapidly to recruitment of monocytes and T-cells at vulnerable sites, setting the stage for inflammation. Then the complex dance of cytokines, growth factors, and immune mechanisms, which we call inflammation, takes off. Many of the inflammatory events, directly or indirectly, may be downstream consequences of the hypercholesterolemia. We shall return to this question below in the discussion of regression. Research on the chronic inflammatory events in atherosclerosis has proceeded at a furious pace over the last decades and several excellent reviews are available (31;107–110). Here it may be helpful to review from an historical point of view some of the key findings that laid the foundation for these new and exciting directions.

Virchow, in 1856, noted that lesion formation began with a thickening of the intima, which he attributed to insudation of blood components into the artery wall. He also noted an early increase in the numbers of subendothelial cells, an increase in their size, and the presence of intracellular lipoid substances, but he had no way to deduce their origin (1).

Over 50 years later, Anitschkow beautifully described the foam cells in the lesions of his cholesterol-fed rabbits and surmised that they represented invading white blood cells (2). He reasoned this way: "one can see quite distinctly all conceivable transitional forms between the small lymphocytic and monocytic cells, on the one hand, and the large lipoid cells on the other." He also reported "lymphoid or monocytic cells frequently seem to invade the aortic wall directly from the lumen." The cells contained lipid globules that were anisotropic under polarized light, indicating the presence of cholesterol esters. Anitschkow favored the view that the foam cells represented mononuclear blood cells, but he had no proof. The origin of the foam cell remained an issue for some time.

In 1958, Poole and Florey (12), using light microscopy, observed lipid-laden macrophages in the subendothelial space of cholesterol-fed rabbits and also on the endothelial surface.

In 1979, Fowler and co-workers developed a method for harvesting foam cells from aortas of cholesterol-fed rabbits so they could be characterized. Using cell-specific monoclonal antibodies, they definitively showed that these cells, from early fatty streak lesions, were mostly of monocyte/macrophage origin (22).

Using electron microscopy, Gerrity in 1981 caught monocytes in the act of penetrating between aortic endothelial cells in cholesterol-fed swine (9;21). More precisely, he saw adherent monocytes with processes extending through narrow gaps but, as he pointed out, there were no arrows to tell whether they were entering or exiting. The former seemed most reasonable. He showed that in the earliest fatty streak lesions the endothelium overlying the foam cells was intact but that in later fatty streaks the endothelium showed gaps, exposing the underlying foam cells.

The mechanism by which the monocyte takes up lipids was elucidated by the Brown/Goldstein laboratory in 1979 (64;71). As discussed earlier in this chapter, they showed that monocyte/macrophages could not be converted to foam cells by incubation with even very high concentrations of native LDL. Uptake was slow, and down regulation of the LDL receptor limited accumulation. On the other hand, macrophages (but not monocytes) expressed a new lipoprotein receptor, dubbed the scavenger receptor, which specifically recognized chemically modified forms of LDL but not native LDL. The modified forms could convert macrophages to foam cells. Evidently the monocyte was undergoing differentiation after entering the lesion site and expressing new sets of receptors. This work, summarized in the 1983 *Annual Review of Biochemistry* (72) attracted the attention of workers in the field and proved to be importantly heuristic.

The discovery of T-lymphocytes in atherosclerotic lesions was first reported from Hansson's laboratory in 1985 (111) and that finding spurred renewed interest in the possible role of the immune system in atherogenesis. The later finding that a subset of T-cells isolated from lesions was specifically activated by oxidized LDL added considerable weight to that possibility (112).

The potential involvement of humoral immunity in atherogenesis was suggested by the 1989 discovery in Witztum's laboratory of the widespread occurrence of autoantibodies against oxidized LDL in the serum of animals and humans (113). Later his laboratory showed that immunization of LDL receptor-deficient rabbits with autologous LDL previously treated with malondialdehyde (a model for oxidized LDL) ameliorated the severity of their atherosclerosis (114). This was the first indication that "vaccination" against atherosclerosis might be a real possibility.

The linkage between monocyte recruitment and hypercholesterolemia was first suggested in the late 1980s by the ability of oxidized LDL to enhance recruitment directly (as a chemoattractant) and/or indirectly (by increasing expression of MCP-1 on endothelial cells) (80;115–118).

In 1991, Cybulsky in Gimbrone's laboratory identified VCAM-1 as an endothelial cell monocyte adhesion molecule (119) and later showed that it was induced soon after feeding rabbits an atherogenic diet, appearing even before foam cells could be found at susceptible arterial sites (120).

Libby and co-workers identified a number of cytokines and growth factors in atherosclerotic lesions or in cultured vascular cells, including MCP-1, FGF, MCSF (121–123). They also described the presence of metalloproteinases in lesions and suggested that their activity could contribute to the thinning of the fibrous cap observed in unstable plaques (124).

The progression of atherosclerotic lesions involves the interactions among the many cell types that characterize the lesion, including mast cells (125), dendritic cells (126), and NK cells (127) in addition to monocytes and T-lymphocytes. The immune system is clearly involved but it is still unclear exactly how, some studies suggesting an anti-atherogenic effect and others a proatherogenic effect. Research is proceeding apace and several excellent reviews are available (31;107–110).

WEIGHING THE RELATIVE IMPORTANCE OF INFLAMMATION AND HYPERLIPIDEMIA

Whether inflammation alone, i.e. *in the absence of some elevation of blood choles-terol*, can ever initiate atherosclerosis is not clear. Various forms of arteritis can of course be generated, but the lesions do not closely resemble those of human atherosclerosis. There is no evidence for "inflammatory atherosclerosis," that is, atherosclerosis in the absence of hypercholesterolemia. Nevertheless, the rate of progression of lesions once established, at any given level of blood cholesterol, can be strongly influenced by the multiple growth factors and cytokines produced within the lesion (or reaching it from elsewhere in the body) and by the immune system. Hopefully, with better understanding of the complex events triggered by the cells in the lesion and by the immune system it may be possible to identify one or more vital steps – Achilles heel steps – at which intervention will significantly slow the growth of lesions and/or stabilize the plaque independently of the blood cholesterol level.

Hypercholesterolemia, on the other hand, can of itself be a sufficient cause of atherosclerosis, as in patients with familial hypercholesterolemia or in the LDL receptor-deficient rabbit, suggesting that the inflammatory changes are downstream consequences of hypercholesterolemia. In this connection, studies on regression induced by simply returning cholesterol levels to normal are instructive.

REGRESSION OF ATHEROSCLEROSIS

If returning elevated cholesterol levels to normal, without attempting to intervene at the level of the arterial wall with anti-inflammatory measures, induces regression of experimental lesions, it would suggest that hypercholesterolemia is the prime mover driving not only lesion initiation, but also lesion progression. It would imply that the pro-inflammatory processes in the vessel wall are not of themselves sufficient to support the growth and development of lesions independently of hypercholesterolemia.

REGRESSION IN ANIMALS

Anitschkow demonstrated that when cholesterol-fed rabbits were returned to a chow diet there was definite but incomplete regression of the aortic lesions (128). Most of the stainable lipid disappeared, as did most of the foam cells and other cellular elements. However, this was a slow process, taking over a year, and it was far from complete. Partly this was because rabbits on the high cholesterol diet store large amounts of cholesterol in tissues other than the arteries; this is slowly released so that the blood cholesterol remains high for some time. Some rabbits actually show some further progression immediately after the cholesterol-rich diet is stopped (2;129).

The extent to which lesions are potentially reversible was dramatically shown in nonhuman primates by Armstrong *et al.* in Connor's laboratory at the University of Iowa (19;130). They studied Rhesus monkeys that had been fed a fat/cholesterol-rich, atherogenic diet for 17 months and had then been returned to a nonatherogenic diet for 40 months (either a cholesterol-free, low-fat diet or a cholesterol-free but corn oil-rich diet). The results in the two regression groups were comparable and were pooled. The blood cholesterol level in the controls was 140 mg/dl and rose to just over 700 mg/dl on the atherogenic diet. Within 60 days of resuming the normal diet the value had dropped back to 138 mg/dl. A control group was sacrificed at 17 months to establish the baseline severity of the lesions and at 17 months the lesions were quite severe. Mean stenosis in the coronary arteries was 53–65 percent and the lesions were loaded with lipid. After 40 months on the cholesterol-free diet mean stenosis had dropped to 14–26 percent (Figure 5.7).

The cross-sectional area of the intima after 17 months on the atherogenic diet was 0.720 mm^2; after 40 months on one of the regression diets it averaged 0.175 mm^2. The cholesterol content of the coronary arteries had increased from a control value of 6.8 mg/g dry weight to 51.2 during the 17 months on the atherogenic diet; after 40 months on the regression diet it had dropped to 18.1 (131). Thus, about 70 percent of the cholesterol accumulated on the atherogenic diet had been mobilized. This result in a nonhuman primate was striking enough to raise hopes that regression might be possible in the human disease.

Recent studies in mouse models show even more strikingly how dramatic regression can be once hypercholesterolemia is corrected. Desurmont *et al.* (132) studied regression of lesions in apoE-deficient mice (17 weeks old) after rapidly returning their cholesterol levels to normal by introducing the human apoE gene using a CMV-adenoviral vector. To get around the problems of antibody production, they used apoE-deficient mice on a nude mouse background. Total cholesterol levels fell from almost 600 mg/dl to less than 100 mg/dl within 21 days after injecting the adenoviral vector and was maintained at that low level for almost 200 days. Proximal aortic lesions in the adenovirus-treated group were less than 10 percent the area of those in untreated apoE-deficient mice of the same age. Moreover, macrophages and foam cells disappeared and there was re-endothelialization of the artery wall. Because apoE appears to affect atherogenesis to some extent by mechanisms not directly related to blood lipid levels, not all of the changes in Desurmont's study may have been due only to lipid lowering.

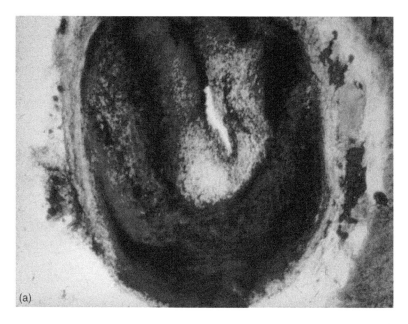

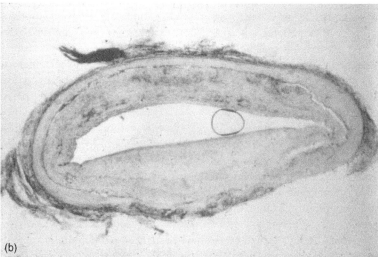

FIGURE 5.7   (a). Almost totally obstructed coronary artery of a Rhesus monkey fed an atherogenic diet by Armstrong *et al.* for 17 months. Blood cholesterol level was about 700 mg/dl. (b). Coronary artery from a Rhesus monkey fed the same atherogenic diet for 17 months but then switched back to a low-fat, low-cholesterol diet for 40 months before sacrifice. Almost all of the stainable lipid has disappeared, the lumen is restored almost to normal and the artery has remodeled considerably. A thick fibrous coat remains. See text for further details of the protocol. Source: photomicrographs courtesy of Dr. William E. Connor. These figures have been reproduced in color in the color plate section.

Fisher's group took a quite different approach (133). They transplanted aortas from 10-month-old apoE-deficient mice either into wild-type mice or, as controls for the surgical procedure, into other apoE-deficient mice. Using this protocol, the atherosclerotic aortic segment is instantly moved from an atherogenic environment to a nonatherogenic environment. Two months after transplantation into wild-type mice, the lesions had all but disappeared; the lesions in aortas transplanted into apoE-deficient recipients had progressed to more than twice the size of those in the pre-transplant aortas.

Taken together, these results suggest that the continuing presence of hypercholesterolemia is needed for lesion maintenance and progression, at least in the case of relatively early lesions. Rather than looking on inflammation and hypercholesterolemia as alternative choices in the pathogenesis of atherosclerosis, it might be more profitable to regard them as "partners in crime," as discussed in detail elsewhere (106).

REGRESSION IN HUMANS

As reviewed by Armstrong (134) and by Wissler and Vesselinovitch (135), several pathologists had reported that the extent and severity of atherosclerosis at autopsy was much less during wartime, when food supplies were limited. The same was true in patients suffering from malnutrition for some time prior to death (e.g. wasting due to cancer or chronic infections such as tuberculosis). The 1947 paper by Wilens on the marked reduction in lesions in terminal cancer cases is particularly persuasive (136). But it was not until the 1970s that direct *in vivo* evidence of anatomic regression began to become available, based on angiography and ultrasound.

In 1972, Knight *et al.* reported angiographic evidence of limited regression of coronary artery lesions in patients whose cholesterol level had been drastically lowered by partial ileal bypass surgery (137). In 1977, Blankenhorn *et al.* published one of the first human studies showing that intensive treatment with diet and drugs could indeed not only arrest progression but actually induce regression in femoral lesions (138).

The widespread acceptance of coronary angiography to evaluate progression or regression of lesions waited on the development of more rigorous approaches. These were systematically developed and exploited by two groups: David H. Blankenhorn and his colleagues at the University of Southern California (139;140); and B. Greg Brown and colleagues at the University of Washington (141;142). Before their work, angiography was more art than science and investigators were skeptical, with good reason, about conclusions drawn from "eyeballing" an angiogram. Their laboratories put quantitative angiography on the map and, down the road, it contributed significantly to the validation of the lipid hypothesis.

STATUS OF THE LIPID HYPOTHESIS IN THE 1980s

The development of a pathogenetic scheme showing how LDL and other proatherogenic lipoproteins penetrate into the artery wall and give rise to foam cells, the hallmark of the initial lesion, together with specific details about the macrophage receptors involved, made it easier to accept the lipid hypothesis. At the same time the new insights into the inflammatory facets of atherogenesis for a time diverted attention from the role of hypercholesterolemia and may have delayed acceptance of the central importance of treating it aggressively. However, even with these advances in understanding, there was little enthusiasm in the early 1980s for lowering blood cholesterol levels as a preventive measure. It would take a definitive large-scale clinical intervention trial to make an airtight case and justify a national preventive program.

NOTE

1   Just as this book was going to press, the clinical trials with torcetrapib were stopped because of toxicity in the treated group. The mechanism underlying the toxicity is under intensive investigation.

REFERENCES

1. Virchow, R. 1856. *Phlogose und Thrombose im Gefass-system*. Meidinger Son and Co., Frankfurt.
2. Anitschkow, N. 1933. Experimental atherosclerosis in animals. In *Arteriosclerosis*, E.V. Cowdry, ed. Macmillan, New York:271–322.
3. Zinserling, W.D. 1925. Untersuchungen uber Atherosklerose. 1. Uber die Aortaverfettung bei Kindern. *Virchow's Archiv f path Anat* 255:676–705.
4. Wissler, R.W. 1995. An overview of the quantitative influence of several risk factors on progression of atherosclerosis in young people in the United States. Pathobiological Determinants of Atherosclerosis in Youth (PDAY) Research Group. *Am J Med Sci* 310 Suppl 1:S29–S36.
5. Long, E.R. 1933. The development of our knowledge of arteriosclerosis. In *Arteriosclerosis; A Survey of the Problem*. E.V. Cowdry, ed. The Macmillan Company, New York:10–52.
6. Ophuls, W. 1933. *The Pathogenesis of Arteriosclerosis*. The Macmillan Company, New York: 249–270.
7. Leary, T. 1934. Experimental atherosclerosis in the rabbit compared with human (coronary) atherosclerosis. *Arch Path* 17:453–492.
8. Leary, T. 1941. The genesis of atherosclerosis. *Arch Path* 32:507–555.
9. Gerrity, R.G. 1981. The role of the moncyte in atherogenesis. II. Migration of foam cells from atherosclerotic lesions. *Am J Pathol* 103:191–200.
10. Duff, G.L. and McMillan, G.C. 1951. Pathology of atherosclerosis. *Am J Med* 11:92–108.
11. Gofman, J.W., Lindgren, F., Elliott, H., Mantz, W., Hewitt, J., and Herring, V. 1950. The role of lipids and lipoproteins in atherosclerosis. *Science* 111:166–171.

12. Poole, J.C.F. and Florey, H.W. 1958. Changes in the endothelium of the aorta and the behavior of macrophages in experimental atheroma of rabbits. *J Pathol* 75:245–252.
13. Jones, R.J. ed. 1963. *Evolution of the Atherosclerotic Plaque*. University of Chicago Press, Chicago and London:1–360.
14. Haust, M.D. and Moore, R.H. 1963. Significance of the smooth muscle cell in atherogenesis. In *Evolution of the Atherosclerotic Plaque*, R.J. Jones, ed. University of Chicago Press, Chicago:51–63.
15. Bierman, E.L., Eisenberg, S., Stein, O., and Stein, Y. 1973. Very low density lipoprotein "remnant" particles: uptake by aortic smooth muscle cells in culture. *Biochim Biophys Acta* 329:163–169.
16. Bierman, E.L., Stein, O., and Stein, Y. 1974. Lipoprotein uptake and metabolism by rat aortic smooth muscle cells in tissue culture. *Circ Res* 35:136–150.
17. Pitas, R.E., Friera, A., McGuire, J., and Dejager, S. 1992. Further characterization of the acetyl LDL (scavenger) receptor expressed by rabbit smooth muscle cells and fibroblasts. *Arterioscler Thromb* 12:1235–1244.
18. Pitas, R.E. 1990. Expression of the acetyl low density lipoprotein receptor by rabbit fibroblasts and smooth muscle cells. Up-regulation by phorbol esters. *J Biol Chem* 265:12722–12727.
19. Armstrong, M.L., Warner, E.D., and Connor, W.E. 1970. Regression of coronary atheromatosis in rhesus monkeys. *Circ Res* 27:59–67.
20. Wissler, R.W. and Vesselinovitch, D. 1990. Can atherosclerotic plaques regress? Anatomic and biochemical evidence from nonhuman animal models. *Am J Cardiol* 65:33F–40F.
21. Gerrity, R.G. 1981. The role of monocyte in atherogenesis. I. Transition of blood-borne monocytes into foam cells in fatty lesions. *Am J Pathol* 103:181–190.
22. Fowler, S., Shio, H., and Haley, N.J. 1979. Characterization of lipid-laden aortic cells from cholesterol-fed rabbits. IV. Investigation of macrophage-like properties of aortic cell populations. *Lab Invest* 41:372–378.
23. Ross, R., Glomset, J., Kariya, B., and Harker, L. 1974. A platelet-dependent serum factor that stimulates the proliferation of arterial smooth muscle cells in vitro. *Proc Natl Acad Sci U.S.A.* 71:1207–1210.
24. Balk, S.D. 1971. Calcium as a regulator of the proliferation of normal, but not of transformed, chicken fibroblasts in a plasma-containing medium. *Proc Natl Acad Sci U.S.A.* 68:271–275.
25. Kohler, N. and Lipton, A. 1974. Platelets as a source of fibroblast growth-promoting activity. *Exp Cell Res* 87:297–301.
26. Ross, R. and Glomset, J.A. 1976. The pathogenesis of atherosclerosis (first of two parts). *N Engl J Med* 295:369–377.
27. Ross, R. and Glomset, J.A. 1976. The pathogenesis of atherosclerosis (second of two parts). *N Engl J Med* 295:420–425.
28. Harker, L.A., Slichter, S.J., Scott, C.R., and Ross, R. 1974. Homocystinemia. Vascular injury and arterial thrombosis. *N Engl J Med* 291:537–543.
29. Harker, L.A., Ross, R., Slichter, S.J., and Scott, C.R. 1976. Homocystine-induced arteriosclerosis. The role of endothelial cell injury and platelet response in its genesis. *J Clin Invest* 58:731–741.
30. Davies, P.F., Reidy, M.A., Goode, T.B., and Bowyer, D.E. 1976. Scanning electron microscopy in the evaluation of endothelial integrity of the fatty lesion in atherosclerosis. *Atherosclerosis* 25:125–130.
31. Ross, R. 1999. Atherosclerosis – an inflammatory disease. *N Engl J Med* 340:115–126.
32. Benditt, E.P. and Benditt, J.M. 1973. Evidence for a monoclonal origin of human atherosclerotic plaques. *Proc Natl Acad Sci U.S.A.* 70:1753–1756.
33. Linder, D. and Gartler, S.M. 1965. Glucose-6-phosphate dehydrogenase mosaicism: utilization as a cell marker in the study of leiomyomas. *Science* 150:67–69.

34. Pittman, R.C., Attie, A.D., Carew, T.E., and Steinberg, D. 1979. Tissue sites of degradation of low density lipoprotein: application of a method for determining the fate of plasma proteins. *Proc Natl Acad Sci U.S.A.* 76:5345–5349.

35. Carew, T.E., Pittman, R.C., Marchand, E.R., and Steinberg, D. 1984. Measurement in vivo of irreversible degradation of low density lipoprotein in the rabbit aorta. Predominance of intimal degradation. *Arteriosclerosis* 4:214–224.

36. Glomset, J.A., Janssen, E.T., Kennedy, R., and Dobbins, J. 1966. Role of plasma lecithin:cholesterol acyltransferase in the metabolism of high density lipoproteins. *J Lipid Res* 7:638–648.

37. Glomset, J.A. 1968. The plasma lecithins:cholesterol acyltransferase reaction. *J Lipid Res* 9:155–167.

38. Sperry, W.M. 1935. Cholesterol esterase in blood. *J Biol Chem* 111:467–478.

39. Glomset, J.A., Parker, F., Tjaden, M., and Williams, R.H. 1962. The esterification in vitro of free cholesterol in human and rat plasma. *Biochim Biophys Acta* 58:398–406.

40. Glomset, J.A. 1962. The mechanism of the plasma cholesterol esterification reaction: plasma fatty acid transferase. *Biochim Biophys Acta* 65:128–135.

41. Murphy, J.R. 1962. Erythrocyte metabolism. IV. Equilibration of cholesterol-4-C-14 between erythrocytes and variously treated sera. *J Lab Clin Med* 60:571–578.

42. Gotto, A.M., Jr., Miller, N.E., and Oliver, M.F. 1978. *High Density Lipoproteins and Atherosclerosis*, Amsterdam/New York, Elsevier:1–225.

43. Pittman, R.C. and Steinberg, D. 1984. Sites and mechanisms of uptake and degradation of high density and low density lipoproteins. *J Lipid Res* 25:1577–1585.

44. Acton, S., Rigotti, A., Landschulz, K.T., Xu, S., Hobbs, H.H., and Krieger, M. 1996. Identification of scavenger receptor SR-B1 as a high density lipoprotein receptor. *Science* 271: 518–520.

45. Barter, P.J. and Rye, K.A. 2006. Relationship between the concentration and antiatherogenic activity of high-density lipoproteins. *Curr Opin Lipidol* 17:399–403.

46. Hughes, S.D., Verstuyft, J., and Rubin, E.M. 1997. HDL deficiency in genetically engineered mice requires elevated LDL to accelerate atherogenesis. *Arterioscler Thromb Vasc Biol* 17: 1725–1729.

47. Navab, M., Fogelman, A.M., Berliner, J.A., Territo, M.C., Demer, L.L., Frank, J.S., Watson, A.D., Edwards, P.A., and Lusis, A.J. 1995. Pathogenesis of atherosclerosis. *Am J Cardiol* 76: 18C–23C.

48. Navab, M., Anantharamaiah, G.M., and Fogelman, A.M. 2005. The role of high-density lipoprotein in inflammation. *Trends Cardiovasc Med* 15:158–161.

49. Navab, M., Anantharamaiah, G.M., Reddy, S.T., Hama, S., Hough, G., Grijalva, V.R., Yu, N., Ansell, B.J., Datta, G., Garber, D.W., Fogelman, A.M. 2005. Apolipoprotein A-I mimetic peptides. *Arterioscler Thromb Vasc Biol* 25:1325–1331.

50. Barter, P.J. and Kastelein, .J. 2006. Targeting cholesteryl ester transfer protein for the prevention and management of cardiovascular disease. *J Am Coll Cardiol* 47:492–499.

51. Brewer, H.B., Jr. 2004. High-density lipoproteins: a new potential therapeutic target for the prevention of cardiovascular disease. *Arterioscler Thromb Vasc Biol* 24:387–391.

52. Goldstein, J.L. and Brown, M.S. 1973. Familial hypercholesterolemia: identification of a defect in the regulation of 3-hydroxy-3-methylglutaryl coenzyme A reductase activity associated with overproduction of cholesterol. *Proc Natl Acad Sci U.S.A.* 70:2804–2808.

53. Brown, M.S. and Goldstein, J.L. 1974. Familial hypercholesterolemia: defective binding of lipoproteins to cultured fibroblasts associated with impaired regulation of 3- hydroxy-3-methylglutaryl coenzyme A reductase activity. *Proc Natl Acad Sci U.S.A.* 71:788–792.

54. Goldstein, J.L. and Brown, M.S. 1977. The low-density lipoprotein pathway and its relation to atherosclerosis. *Annu Rev Biochem* 46:897–930.

55. Brown, M.S., Dana, S.E., Dietschy, J.M., and Siperstein, M.D. 1973. 3-Hydroxy-3-methylglutaryl coenzyme A reductase. Solubilization and purification of a cold-sensitive microsomal enzyme. *J Biol Chem* 248:4731–4738.

56. Wilkinson, C.F., Hand. E.A., and Fliegelman, M.T. 1948. Essential familial hypercholesterolemia. *Ann Intern Med* 29:671–686.
57. Adlersberg, D., Parets, A.D., and Boas, E.P. 1949. Genetics of atherosclerosis. *JAMA* 141: 246–254.
58. Khachadurian, A.K. 1964. The inheritance of essential familial hypercholesterolemia. *Am J Med* 37:402–407.
59. Gould, R.G. and Taylor, C.B. 1950. Effect of dietary cholesterol on hepatic cholesterol biosynthesis. *Fed Proc* 9:179.
60. Gould, R.G. and Popjak, G. 1957. Biosynthesis of cholesterol in vivo and in vitro from DL-beta-hydroxy-beta-methyl-delta [2-14C]-valerolactone. *Biochem J* 66:51P.
61. Brown, M.S. and Goldstein, J.L. 1976. Receptor-mediated control of cholesterol metabolism. *Science* 191:150–154.
62. Mahley, R.W., Innerarity, T.L., Rall, S.C., Jr., and Weisgraber, K.H. 1984. Plasma lipoproteins: apolipoprotein structure and function. *J Lipid Res* 25:1277–1294.
63. Mahley, R.W., Innerarity, T.L., Brown, M.S., Ho, Y.K., and Goldstein, J.L. 1980. Cholesteryl ester synthesis in macrophages: stimulation by beta- very low density lipoproteins from cholesterol-fed animals of several species. *J Lipid Res* 21:970–980.
64. Goldstein, J.L., Ho, Y.K., Brown, M.S., Innerarity, T.L., and Mahley, R.W. 1980. Cholesteryl ester accumulation in macrophages resulting from receptor-mediated uptake and degradation of hypercholesterolemic canine beta-very low density lipoproteins. *J Biol Chem* 255:1839–1848.
65. Vega, G.L. and Grundy, S.M. 1986. In vivo evidence for reduced binding of low density lipoproteins to receptors as a cause of primary moderate hypercholesterolemia. *J Clin Invest* 78:1410–1414.
66. Innerarity, T.L., Weisgraber, K.H., Arnold, K.S., Mahley, R.W., Krauss, R.M., Vega, G.L., and Grundy, S.M. 1987. Familial defective apolipoprotein B-100: low density lipoproteins with abnormal receptor binding. *Proc Natl Acad Sci U.S.A.* 84:6919–6923.
67. Myant, N.B. 1993. Familial defective apolipoprotein B-100: a review, including some comparisons with familial hypercholesterolaemia. *Atherosclerosis* 104:1–18.
68. Stein, O., Weinstein, D.B., Stein, Y., and Steinberg, D. 1976. Binding, internalization, and degradation of low density lipoprotein by normal human fibroblasts and by fibroblasts from a case of homozygous familial hypercholesterolemia. *Proc Natl Acad Sci U.S.A.* 73:14–18.
69. Lehrman, M.A., Goldstein, J.L., Brown, M.S., Russell, D.W., and Schneider, W.J. 1985. Internalization-defective LDL receptors produced by genes with nonsense and frameshift mutations that truncate the cytoplasmic domain. *Cell* 41:735–743.
70. Brown, M.S., Goldstein, J.L., Krieger, M., Ho, Y.K., and Anderson, R.G. 1979. Reversible accumulation of cholesteryl esters in macrophages incubated with acetylated lipoproteins. *J Cell Biol* 82:597–613.
71. Goldstein, J.L., Ho, Y.K., Basu, S.K., and Brown, M.S. 1979. Binding site on macrophages that mediate uptake and degradation of acetylated low density lipoprotein, producing massive cholesterol deposition. *Proc Natl Acad Sci U.S.A.* 76:333–337.
72. Brown, M.S. and Goldstein, J.L. 1983. Lipoprotein metabolism in the macrophage: implications for cholesterol deposition in atherosclerosis. *Annu Rev Biochem* 52:223–261.
73. Fogelman, A.M., Shechter, I., Seager, J., Hokom, M., Child, J.S., and Edwards, P.A. 1980. Malondialdehyde alteration of low density lipoproteins leads to cholesteryl ester accumulation in human monocyte-macrophages. *Proc Natl Acad Sci U.S.A.* 77:2214–2218.
74. Mahley, R.W., Innerarity, T.L., Weisgraber, K.H., and Oh, S. 1979. Altered metabolism (in vivo and in vitro) of plasma lipoproteins after selective chemical modification of lysine residues of the apoproteins. *J Clin Invest* 64:743–750.
75. Henriksen, T., Evensen, S., and Carlander, B. 1979. Injury to human endothelial cells in culture induced by low density lipoproteins. *Scand J Clin Lab Invest* 39:361–368.

76. Hessler, J.R., Robertson, A.L., Jr., and Chisolm, G.M. 1979. LDL-induced cytotoxicity and its inhibition by HDL in human vascular smooth muscle and endothelial cells in culture. *Atherosclerosis* 32:213–229.

77. Henriksen, T., Mahoney, E.M., and Steinberg, D. 1981. Enhanced macrophage degradation of low density lipoprotein previously incubated with cultured endothelial cells: recognition by receptors for acetylated low density lipoproteins. *Proc Natl Acad Sci U.S.A.* 78:6499–6503.

78. Hessler, J.R., Morel, D.W., Lewis, L.J., and Chisolm, G.M. 1983. Lipoprotein oxidation and lipoprotein-induced cytotoxicity. *Arteriosclerosis* 3:215–222.

79. Steinbrecher, U.P., Parthasarathy, S., Leake, D.S., Witztum, J.L., and Steinberg, D. 1984. Modification of low density lipoprotein by endothelial cells involves lipid peroxidation and degradation of low density lipoprotein phospholipids. *Proc Natl Acad Sci U.S.A.* 81:3883–3887.

80. Quinn, M.T., Parthasarathy, S., Fong, L.G., and Steinberg, D. 1987. Oxidatively modified low density lipoproteins: a potential role in recruitment and retention of monocyte/macrophages during atherogenesis. *Proc Natl Acad Sci U.S.A.* 84:2995–2998.

81. Naruszewicz, M., Carew, T.E., Pittman, R.C., Witztum, J.L., and Steinberg, D. 1984. A novel mechanism by which probucol lowers low density lipoprotein levels demonstrated in the LDL receptor-deficient rabbit. *J Lipid Res* 25:1206–1213.

82. Parthasarathy, S., Young, S.G., Witztum, J.L., Pittman, R.C., and Steinberg, D. 1986. Probucol inhibits oxidative modification of low density lipoprotein. *J Clin Invest* 77:641–644.

83. Carew, T.E., Schwenke, D.C., and Steinberg, D. 1987. Antiatherogenic effect of probucol unrelated to its hypocholesterolemic effect: evidence that antioxidants in vivo can selectively inhibit low density lipoprotein degradation in macrophage-rich fatty streaks and slow the progression of atherosclerosis in the Watanabe heritable hyperlipidemic rabbit. *Proc Natl Acad Sci U.S.A.* 84:7725–7729.

84. Kita, T., Nagano, Y., Yokode, M., Ishi, K., Kume, N., Ooshima, A., Yoshida, H., and Kawai, C. 1987. Probucol prevents the progression of atherosclerosis in Watanabe heritable hyperlipidemic rabbit, an animal model for familial hypercholesterolemia. *Proc Natl Acad Sci U.S.A.* 84:5928–5931.

85. Wu, B.J., Kathir, K., Witting, P.K., Beck, K., Choy, K., Li, C., Croft, K.D., Mori, T.A., Tanous, D., Adams, M.R. Lau, A.K., and Stocker, R. 2006. Antioxidants protect from atherosclerosis by a heme oxygenase-1 pathway that is independent of free radical scavenging. *J Exp Med* 203:1117–1127.

86. Steinberg, D., Parthasarathy, S., Carew, T.E., Khoo, J.C., and Witztum, J.L. 1989. Beyond cholesterol. Modifications of low-density lipoprotein that increase its atherogenicity. *N Engl J Med* 320:915–924.

87. Navab, M., Berliner, J.A., Watson, A.D., Hama, S.Y., Territo, M.C., Lusis, A.J., Shih, D.M., Van Lenten, B.J., Frank, J.S., Demer, L.L., Edwards, P.A., and Fogelman, A.M. 1996. The Yin and Yang of oxidation in the development of the fatty streak. A review based on the 1994 George Lyman Duff Memorial Lecture. *Arterioscler Thromb Vasc Biol* 16:831–842.

88. Chisolm, G.M. and Steinberg, D. 2000. The oxidative modification hypothesis of atherogenesis: an overview. *Free Radic Biol Med* 28:1815–1826.

89. Berliner, J.A. and Heinecke, J.W. 1996. The role of oxidized lipoproteins in atherogenesis. *Free Radic Biol Med* 20:707–727.

90. Steinberg, D. and Witztum, J.L. 2002. Is the oxidative modification hypothesis relevant to human atherosclerosis? Do the antioxidant trials conducted to date refute the hypothesis? *Circulation* 105:2107–2111.

91. Steinberg, D. 1992. Antioxidants in the prevention of human atherosclerosis. Summary of the proceedings of a National Heart, Lung, and Blood Institute Workshop: September 5–6, 1991, Bethesda, Maryland. *Circulation* 85:2337–2344.

92. Heinecke, J.W. 2001. Is the emperor wearing clothes? Clinical trials of vitamin E and the LDL oxidation hypothesis. *Arterioscler Thromb Vasc Biol* 21:1261–1264.

93. Vivekananthan, D.P., Penn, M.S., Sapp, S.K., Hsu, A., and Topol, E.J. 2003. Use of antioxidant vitamins for the prevention of cardiovascular disease: meta-analysis of randomised trials. *Lancet* 361:2017–2023.

94. Khoo, J.C., Miller, E., McLoughlin, P., and Steinberg, D. 1988. Enhanced macrophage uptake of low density lipoprotein after self-aggregation. *Arteriosclerosis* 8:348–358.

95. Frank, J.S. and Fogelman, A.M. 1989. Ultrastructure of the intima in WHHL and cholesterol-fed rabbit aortas prepared by ultra-rapid freezing and freeze-etching. *J Lipid Res* 30:967–978.

96. Camejo, G., Hurt-Camejo, E., Wiklund, O., and Bondjers, G. 1998. Association of apo B lipoproteins with arterial proteoglycans: pathological significance and molecular basis. *Atherosclerosis* 139:205–222.

97. Khoo, J.C., Miller, E., Pio, F., Steinberg, D., and Witztum, J.L. 1992. Monoclonal antibodies against LDL further enhance macrophage uptake of LDL aggregates. *Arterioscler Thromb* 12:1258–1266.

98. Kodama, T., Reddy, P., Kishimoto, C., and Krieger, M. 1988. Purification and characterization of a bovine acetyl low density lipoprotein receptor. *Proc Natl Acad Sci U.S.A.* 85:9238–9242.

99. Sparrow, C.P., Parthasarathy, S., and Steinberg, D. 1989. A macrophage receptor that recognizes oxidized low density lipoprotein but not acetylated low density lipoprotein. *J Biol Chem* 264:2599–2604.

100. Endemann, G., Stanton, L.W., Madden, K.S., Bryant, C.M., White, R.T., and Protter, A.A. 1993. CD36 is a receptor for oxidized low density lipoprotein. *J Biol Chem* 268: 11811–11816.

101. Nicholson, A.C., Frieda, S., Pearce, A., and Silverstein, R.L. 1995. Oxidized LDL binds to CD36 on human monocyte-derived macrophages and transfected cell lines. Evidence implicating the lipid moiety of the lipoprotein as the binding site. *Arterioscler Thromb Vasc Biol* 15:269–275.

102. Febbraio, M., Podrez, E.A., Smith, J.D., Hajjar, D.P., Hazen, S.L., Hoff, H.F., Sharma,K., and Silverstein, R.L. 2000. Targeted disruption of the class B scavenger receptor CD36 protects against atherosclerotic lesion development in mice. *J Clin Invest* 105:1049–1056.

103. Suzuki, H., Kurihara, Y., Takeya, M., Kamada, N., Kataoka, M., Jishage, K., Ueda, O., Sakaguchi, H., Higashi, T., Suzuki, T., Takashima, Y., Kawabe, Y., Cynshi, O., Wada, Y., Honda, M., Kurihara, H., Aburatani, H., Doi, T., Matsumoto, A., Azuma, S., Noda, T., Toyoda, Y., Itakura, H., Yazaki, Y., Hriuchi, S., Takahashi, K., Kruijt, J.K., vanBerkel, T.J.C., Steinbrecher, U.P., Ishibashi, S., Maeda, N., Gordon, S., and Kodama, T. 1997. A role for macrophage scavenger receptors in atherosclerosis and susceptibility to infection. *Nature* 386:292–296.

103A. Febbraio, M., Guy, E., and Silverstein, R.L. 2004. Stem cell transplantation reveals that absence of macrophage CD36 is protective against atherosclerosis. Arterioscler Thromn Vasc Biol 24:2333–2338.

103B. Moore, K.J., Kunjathoor, V.V., Koehn, S.L., Manning, J.J., Tseng, A.A., Silver, J.M., McKee, M., and Freeman, M.W. 2005 Loss of receptor-mediated uptake via scavenger receptor A or CD36 pathways does not ameliorate atherosclerosis in hyperlipidemic mice. *J Clin Invest* 115:2192–2201.

103C. Witztum, J.L. 2005. You are right too! J Clin Invest 115:2072–2075

104. Ross, R. and Glomset, J.A. 1973. Atherosclerosis and the arterial smooth muscle cell: proliferation of smooth muscle is a key event in the genesis of the lesions of atherosclerosis. *Science* 180:1332–1339.

105. Ross, R. 1993. The pathogenesis of atherosclerosis: a perspective for the 1990s. *Nature* 362:801–809.

106. Steinberg, D. 2002. Atherogenesis in perspective: hypercholesterolemia and inflammation as partners in crime. *Nat Med* 8:1211–1217.
107. Binder, C.J., Chang, M.K., Shaw, P.X., Miller, Y.I., Hartvigsen, K., Dewan, A., and Witztum, J.L. 2002. Innate and acquired immunity in atherogenesis. *Nat Med* 8:1218–1226.
108. Libby, P., Hansson, G.K., and Pober, J.S. 1999. Atherogenesis and inflammation. In *Molecular Basis of Cardiovascular Disease*, Chien, K.R., ed. Philadelphia, W.B. Saunders: 349–366.
109. Hansson, G.K. 2001. Immune mechanisms in atherosclerosis. *Arterioscler Thromb Vasc Biol* 21:1876–1890.
110. Getz, G.S. 2005. Thematic review series: the immune system and atherogenesis. Immune function in atherogenesis. *J Lipid Res* 46:1–10.
111. Jonasson, L., Holm, J., Skalli, O., Gabbiani, G., and Hansson, G.K. 1985. Expression of class II transplantation antigen on vascular smooth muscle cells in human atherosclerosis. *J Clin Invest* 76:125–131.
112. Stemme, S., Faber, B., Holm, J., Wiklund, O., Witztum, J.L., and Hansson, G.K. 1995. T lymphocytes from human atherosclerotic plaques recognize oxidized low density lipoprotein. *Proc Natl Acad Sci U.S.A.* 92:3893–3897.
113. Palinski, W., Rosenfeld, M.E., Ylä-Herttuala, S., Gurtner, G.C., Socher, S.S., Butler, S.W., Parthasarathy, S., Carew, T.E., Steinberg, D., and Witztum, J.L. 1989. Low density lipoprotein undergoes oxidative modification in vivo. *Proc Natl Acad Sci U.S.A.* 86:1372–1376.
114. Palinski, W., Miller, E., and Witztum, J.L. 1995. Immunization of low density lipoprotein (LDL) receptor-deficient rabbits with homologous malondialdehyde-modified LDL reduces atherogenesis. *Proc Natl Acad Sci U.S.A.* 92:821–825.
115. Quinn, M.T., Parthasarathy, S., and Steinberg, D. 1985. Endothelial cell-derived chemotactic activity for mouse peritoneal macrophages and the effects of modified forms of low density lipoprotein. *Proc Natl Acad Sci U.S.A.* 82:5949–5953.
116. Quinn, M.T., Parthasarathy, S., and Steinberg, D. 1988. Lysophosphatidylcholine: a chemotactic factor for human monocytes and its potential role in atherogenesis. *Proc Natl Acad Sci U.S.A.* 85:2805–2809.
117. Berliner, J.A., Territo, M.C., Sevanian, A., Ramin, S., Kim, J.A., Bamshad, B., Esterson, M., and Fogelman, A.M. 1990. Minimally modified low density lipoprotein stimulates monocyte endothelial interactions. *J Clin Invest* 85:1260–1266.
118. Cushing, S.D., Berliner, J.A., Valente, A.J., Territo, M.C., Navab, M., Parhami, F., Gerrity, R., Schwartz, C.J., and Fogelman, A.M. 1990. Minimally modified low density lipoprotein induces monocyte chemotactic protein 1 in human endothelial cells and smooth muscle cells. *Proc Natl Acad Sci U.S.A.* 87:5134–5138.
119. Cybulsky, M.I. and Gimbrone, M.A., Jr. 1991. Endothelial expression of a mononuclear leukocyte adhesion molecule during atherogenesis. *Science* 251:788–791.
120. Li, H., Cybulsky, M.I., Gimbrone, M.A., Jr., and Libby, P. 1993. An atherogenic diet rapidly induces VCAM-1, a cytokine-regulatable mononuclear leukocyte adhesion molecule, in rabbit aortic endothelium. *Arterioscler Thromb* 13:197–204.
121. Clinton, S.K., Underwood, R., Hayes, L., Sherman, M.L., Kufe, D.W., and Libby, P. 1992. Macrophage colony-stimulating factor gene expression in vascular cells and in experimental and human atherosclerosis. *Am J Pathol* 140:301–316.
122. Brogi, E., Winkles, J.A., Underwood, R., Clinton, S.K., Alberts, G.F., and Libby, P. 1993. Distinct patterns of expression of fibroblast growth factors and their receptors in human atheroma and nonatherosclerotic arteries. Association of acidic FGF with plaque microvessels and macrophages. *J Clin Invest* 92:2408–2418.
123. Wang, J.M., Sica, A., Peri, G., Walter, S., Padura, I.M., Libby, P., Ceska, M., Lindley, I., Colotta, F., and Mantovani, A. 1991. Expression of monocyte chemotactic protein and

interleukin-8 by cytokine-activated human vascular smooth muscle cells. *Arterioscler Thromb* 11:1166–1174.

124. Galis, Z.S., Sukhova, G.K., Lark, M.W., and Libby, P. 1994. Increased expression of matrix metalloproteinases and matrix degrading activity in vulnerable regions of human atherosclerotic plaques. *J Clin Invest* 94:2493–2503.

125. Lindstedt, K.A. and Kovanen, P.T. 2004. Mast cells in vulnerable coronary plaques: potential mechanisms linking mast cell activation to plaque erosion and rupture. *Curr Opin Lipidol* 15:567–573.

126. Bobryshev, Y.V. and Lord, R.S. 1995. S-100 positive cells in human arterial intima and in atherosclerotic lesions. *Cardiovasc Res* 29:689–696.

127. Bobryshev, Y.V. and Lord, R.S. 2005. Identification of natural killer cells in human atherosclerotic plaque. *Atherosclerosis* 180:423–427.

128. Anitschkow, N.N. 1928. Ueber die Ruckbildungsvorgange bei der experimentellen Atherosklerose. *Verhandlungen der Deutschen Pathologischen Gesellschaft* 23rd Conference:473–478.

129. Constantinides, P. 1965. *Experimental Atherosclerosis.* Elsevier, Amsterdam/London/New York:1–93.

130. Armstrong, M.L. 1977. Connective tissue changes in regression. In *Atherosclerosis IV, Proceedings of the Fourth International Symposium*, G. Schettler, Y. Goto, Y. Hata, and G. Klose, eds. Springer-Verlag, Berlin:405–413.

131. Armstrong, M.L. and Megan, M.B. 1972. Lipid depletion in atheromatous coronary arteries in rhesus monkeys after regression diets. *Circ Res* 30:675–680.

132. Desurmont, C., Caillaud, J.M., Emmanuel, F., Benoit, P., Fruchart, J.C., Castro, G., Branellec, D., Heard, J.M., and Duverger, N. 2000. Complete atherosclerosis regression after human ApoE gene transfer in ApoE-deficient/nude mice. *Arterioscler Thromb Vasc Biol* 20:435–442.

133. Reis, E.D., Li, J., Fayad, Z.A., Rong, J.X., Hansoty, D., Aguinaldo, J.G., Fallon, J.T., and Fisher, E.A. 2001. Dramatic remodeling of advanced atherosclerotic plaques of the apolipoprotein E-deficient mouse in a novel transplantation model. *J Vasc Surg* 34:541–547.

134. Armstrong, M.L. 1976. Evidence of regression of atherosclerosis in primates and man. *Postgrad Med J* 52:456–461.

135. Wissler, R.W. and Vesselinovitch, D. 1977. Regression of atherosclerosis in experimental animals and man. *Mod Concepts Cardiovasc Dis* 46:27–32.

136. Wilens, S.L. 1947. The resorption of arterial atheromatous deposits in wasting disease. *Am J Pathol* 23:793–804.

137. Knight, L., Scheibel, R., Amplatz, K., Varco, R.L., and Buchwald, H. 1972. Radiographic appraisal of the Minnesota partial ileal bypass study. *Surg Forum* 23:141–142.

138. Barndt, R., Jr., Blankenhorn, D.H., Crawford, D.W., and Brooks, S.H. 1977. Regression and progression of early femoral atherosclerosis in treated hyperlipoproteinemic patients. *Ann Intern Med* 86:139–146.

139. Crawford, D.W., Beckenbach, E.S., Blankenhorn, D.H., Selzer, R.H., and Brooks, S.H. 1974. Grading of coronary atherosclerosis. Comparison of a modified IAP visual grading method and a new quantitative angiographic technique. *Atherosclerosis* 19:231–241.

140. Blankenhorn, D.H. and Hodis, H.N. 1994. George Lyman Duff Memorial Lecture. Arterial imaging and atherosclerosis reversal. *Arterioscler Thromb* 14:177–192.

141. Brown, B.G., Bolson, E., Frimer, M., and Dodge, H.T. 1977. Quantitative coronary arteriography: estimation of dimensions, hemodynamic resistance, and atheroma mass of coronary artery lesions using the arteriogram and digital computation. *Circulation* 55:329–337.

142. Brown, G., Albers, J.J., Fisher, L.D., Schaefer, S.M., Lin, J.T., Kaplan, C., Zhao, X.Q., Bisson, B.D., Fitzpatrick, V.F., and Dodge, H.T. 1990. Regression of coronary artery disease as a result of intensive lipid-lowering therapy in men with high levels of apolipoprotein B. *N Engl J Med* 323:1289–1298.

CHAPTER 6

The Search for Cholesterol-lowering Drugs

By 1970, the lipid hypothesis was strongly supported by animal model data, by genetic evidence, by epidemiologic correlations, and by several large-scale clinical trials using cholesterol-lowering diets, as reviewed in the preceding chapters. Nevertheless, the hypothesis was still widely regarded with skepticism, at best, or summarily dismissed, at worst. Certainly little or nothing was being done yet at the clinical level to reduce blood cholesterol levels. Furthermore, those few practitioners who were already persuaded that lowering cholesterol levels would be beneficial were frustrated because they simply did not have effective tools for lowering it. Diets very low in saturated fat were certainly effective under controlled metabolic ward conditions (1;2). However, the dietary modifications required for optimal effects were unacceptable to most people in our fast food society. Moreover, even with good compliance, dietary intervention was often without much effect in the very patients who needed treatment the most – those with severe hypercholesterolemia due to genetically determined dyslipidemias. Consequently the hunt for drugs that might safely and effectively reduce blood cholesterol continued.

HYPOCHOLESTEROLEMIC DRUGS AS A TARGET: CONS AND PROS

Looking for new cholesterol-lowering drugs was hardly a high priority for the pharmaceutical industry in the 1960s and 1970s. The lipid hypothesis was not widely accepted and practitioners just didn't think prevention was likely to work, certainly not by just lowering blood cholesterol a bit. Taking a new drug through from bench to clinical trial was a very expensive proposition. Would there be a market at the other end?

Drugs lowering cholesterol levels would have to be taken lifelong and thus the safety margin would have to be extraordinary. As discussed below, the European WHO clofibrate trial, while showing a significant reduction in nonfatal myocardial infarction, showed a significant increase in overall mortality (3). Moreover,

The Cholesterol Wars by D. Steinberg.
Copyright © 2007 Elsevier Inc. All rights of reproduction in any form reserved.

also discussed below, three of the four drugs used in the Coronary Drug Project in the U.S. – estrogenic hormone, clofibrate, and dextrothyroxine – proved to be toxic or ineffective or both (4). Understandably, drug discovery teams felt there were more fertile fields to till.

On the other hand, with coronary heart disease the major cause of death, the potential to have an impact and at the same time to have a healthy bottom line was great. The requirement for lifelong treatment, while a hurdle from the point of view of drug safety, insured sustained profits if one got lucky. It was a tough call and different companies responded differently.

In this chapter we analyze the histories of the drugs that made it through to clinical use, excepting drugs inhibiting cholesterol biosynthesis, which are taken up in Chapter 8 along with the statins.

NICOTINIC ACID

The first clinically effective cholesterol-lowering drug was nicotinic acid. It was discovered by Rudolf Altschul (5). The research leading to the discovery was based on a plausible but, in retrospect, quite incorrect hypothesis, a not uncommon occurrence in science. However, whatever the mechanism of action, nicotinic acid did work and it remains to this day a widely used and effective treatment, albeit not without some toxic side effects.

ALTSCHUL'S RATIONALE

Rudolf Altschul was born in 1901 in Prague where he got his medical training. After postdoctoral studies in Paris and Rome he set up a practice of neuropsychiatry in Prague but in 1939 was forced to flee to Canada after the Nazi occupation. His scientific interests were very broad, and during his tenure as Professor of Anatomy at the University of Saskatchewan he explored a number of problems in the pathology of the nervous system, muscular degeneration, and arterial degeneration. For the latter studies he used the cholesterol-fed rabbit model of Anitschkow. One of his observations was that if the cholesterol used in the feeding studies had been previously exposed to ultraviolet irradiation, the severity of the lesions was reduced (6). This led him to try exposing rabbits on a diet rich in untreated cholesterol to ultraviolet light, instead of irradiating the dietary cholesterol itself. The rabbits were shaved and exposed to ultraviolet light three times a week for 20 to 60 minutes over a total of 90 days of cholesterol feeding. He reported in 1953 that this ultraviolet treatment markedly reduced the levels of hypercholesterolemia in the rabbits and also the severity of the atherosclerotic lesions (7). Pursuing the hypothesis that some sort of oxidative process was responsible for the effect, he tried exposing cholesterol-fed rabbits to oxygen and

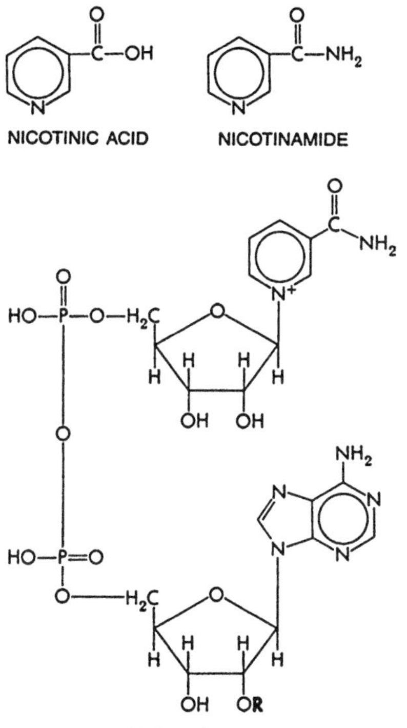

FIGURE 6.1   Structures of nicotinic acid, nicotinamide, and nicotinamide adenine dinucleotide (NAD).

found that this also ameliorated the degree of hypercholesterolemia and the severity of the lesions (8). Attributing the findings in both cases to an increase in cholesterol oxidation, he decided to try increasing endogenous cholesterol oxidation in the cholesterol-fed rabbit by feeding nicotinic acid, a major component of the respiratory coenzyme nicotinamide adenine dinucleotide (NAD), see Figure 6.1.

Nicotinic acid worked, apparently supporting the hypothesis. However, it only worked if given in huge doses, doses three orders of magnitude greater than the doses needed for maximizing tissue concentrations of the cofactor – grams instead of milligrams (5;9). The other problem with the hypothesis, as Altschul himself recognized early on, was that the amide form of nicotinic acid had no effect whatever on cholesterol levels even when fed in the same gram quantities as the acid. Since the two forms are equivalent in providing precursor for endogenous NAD production, the ineffectiveness of the amide form as a cholesterol-lowering agent pretty much ruled out the original hypothesis. Whatever the mechanism of action, the treatment was effective in lowering cholesterol levels in humans and this was quickly confirmed (10–12). Many years later it was chosen as one of the cholesterol-lowering drugs tested for efficacy in preventing coronary events in the Coronary Drug Project (13).

HOW NICOTINIC ACID REALLY WORKS AND HOW EFFECTIVE IT REALLY IS!

We know now that the mechanism of action of nicotinic acid rests largely on its ability to inhibit the release of free fatty acids from adipose tissue, as shown first by Carlson and co-workers (14;15), and has nothing to do with cholesterol oxidation as originally postulated by Altschul. The rate of delivery of free fatty acids to the liver is a major determinant of the rate of hepatic production of VLDL, the precursor of LDL, and thus nicotinic acid treatment decreases the plasma levels of both triglycerides and cholesterol. In addition it *raises* HDL levels, an effect that would be expected to further reduce coronary heart disease risk. Indeed, nicotinic acid treatment as monotherapy has been shown to reduce rates of myocardial infarction, both fatal and nonfatal. During the 6.5 years of the Coronary Drug Project (16), the men on nicotinic acid (3 g/d) had a 27 percent reduction in nonfatal myocardial infarction, a 26 percent reduction in cerebrovascular events, and a 15 percent reduction in the sum of nonfatal myocardial infarction and death. There was no decrease in all-cause mortality during the 6.5 years of the trial but a follow-up on the cohort at 15 years showed a statistically significant 11 percent decrease in all-cause mortality (17). Thus nicotinic acid has the distinction of being the first cholesterol-lowering drug shown to decrease cardiovascular events, cardiovascular mortality, and all-cause mortality.

Recently the mechanism of action has been further clarified by the discovery that nicotinic acid is a specific agonist for a G_i-protein-coupled receptor, GPR109A (18–20). Both the suppression of free fatty acid release and the flushing reaction are mediated via this receptor, the latter effect being mediated by downstream effects on cyclooxygenase and prostaglandin production. The physiologic function of this receptor remains uncertain but there is no doubt that it mediates the pharmacologic effects of nicotinic acid.

So, some 50 years after Altschul stumbled on the effectiveness of high doses of nicotinic acid in lowering plasma cholesterol levels, we now understand its mode of action at a molecular level.

SIDE EFFECTS

While clinically effective, nicotinic acid had side effects that made it unacceptable to many patients, mainly the flushing and itching that follow soon after taking the drug. The severity of these symptoms varied considerably from patient to patient, and it could be ameliorated by starting at very low dosages (100 mg/d) and titrating up gradually to the clinically effective steady-state dosages of 3–6 g/d. Taking an aspirin tablet 30–60 minutes before the nicotinic acid also helped (as we know now, by inhibiting prostaglandin release). Still, many physicians would not risk alienating their patients by prescribing nicotinic acid. Another reason it failed to find favor is that it impaired hepatic function in some patients, especially diabetic patients. Slow-release forms of nicotinic acid were developed to try to get around the flushing and itching and this seemed to improve matters. However, the original slow-release preparations, for reasons still not fully understood, caused a number of cases of catastrophic liver failure, some fatal. More recently, a novel formulation (extended release) in which the nicotinic acid is embedded in a wax matrix (Niaspan®) seems to have solved the problem, and the drug is now being much more widely used (21). Lars A. Carlson, whose pioneering work over 40 years ago first showed that nicotinic acid worked by suppressing free fatty acid release, has recently written an authoritative review (22).

BILE ACID-BINDING RESINS

The possibility that inhibition of cholesterol absorption might reduce plasma cholesterol levels and thus inhibit atherogenesis was first suggested by Siperstein, Nichols, and Chaikoff (23). I.L. Chaikoff at Berkeley was one of the first physiologists to exploit fully ^{14}C in the study of metabolism. In 1952, his group demonstrated that when cholesterol-4-^{14}C was fed to rats after tying off the bile duct, there was essentially no label recovered in the thoracic duct lymph (24). Earlier work had already supported the suggestion that bile was needed for the absorption of cholesterol (25) but Chaikoff's use of the highly sensitive radioisotope technique settled the issue: the presence of bile was actually obligatory. It was known that ferric chloride would precipitate bile salts *in vitro* and it was known that feeding cholesterol to cockerels raised their plasma cholesterol and induced atherosclerotic lesions. So the investigators fed one group of cockerels cholesterol alone and a second group cholesterol plus ferric chloride. In the latter group the plasma cholesterol levels reached were much lower and there were many fewer atherosclerotic lesions (26). However, ferric chloride itself was quite toxic and definitely not a viable candidate for clinical use. Nevertheless, this pioneering work set the stage for the later development of the nonabsorbable bile acid-binding resins, which proved to be very effective.

In 1959–61 a group of investigators at the Merck Institute for Therapeutic Research had the clever idea of orally administering a nonabsorbable ion exchange resin to trap bile acids so they would be simply carried out in the feces (27–29). The resin they developed had a styrene-divinylbenzene skeleton to which were attached quaternary ammonium anion exchanging groups. The affinity of the anion exchange groups for bile acids was remarkable: addition of 1 g of the resin to 100 ml of a 1 percent solution of cholic acid (1 g) caused adsorption of over 90 percent of the bile acid. Since the resin was a very high molecular weight solid, there was no intestinal absorption of it and thus no chance of systemic toxicity. Indeed, long-term studies in cockerels and dogs showed no toxic effects. Fat absorption was unimpaired as was vitamin K absorption and storage. In humans treated with cholestyramine, total cholesterol was reduced by 20–25 percent and LDL by 30–35 percent (30). In many ways, therefore, this would seem to be an ideal drug for treatment of hypercholesterolemia. The catch of course is that it comes as a dry, sandy powder that has to be mixed with water or juice and chugalugged two or three times a day. Furthermore, many patients experience unacceptable gastrointestinal side effects (gastritis, bloating, diarrhea, or constipation). Understandably compliance is poor and most physicians shied away from even discussing it with their patients. Despite these considerable disadvantages, cholestyramine was the drug chosen for the critical test of the lipid hypothesis in the later Coronary Primary Prevention Study because the subjects would be getting constant re-enforcement from doctors and staff. While compliance was indeed not the best, it was sufficient to yield an average 11 percent drop in cholesterol level and a statistically significant 19 percent drop in coronary events (31;32).

The history of cholestyramine and the Food and Drug Administration (FDA) sheds light on the evolution of thinking at the agency regarding the cholesterol–coronary heart disease connection. When cholestyramine was first cleared for clinical use in the early 1960s, the approved indications did not include either hypercholesterolemia or coronary heart disease. It was approved for use in the amelioration of pruritis in patients with primary biliary cirrhosis (33)! At that time the FDA still did not consider the cholesterol–coronary heart disease linkage strong enough to warrant health claims for cholesterol-lowering per se. It was not until 1985, after the 1984 report of the NIH-sponsored Coronary Primary Prevention Trial (CPPT), that the agency added lowering of LDL cholesterol as an approved indication for cholestyramine (34). At that time, clofibrate had been approved for lowering LDL-C in patients with Type III hyperlipoproteinemia but approved in patients with Types IV and V *only for prevention of pancreatitis*, by lowering triglycerides. Gemfibrozil also was approved but only in patients with high triglyceride levels and *at risk for pancreatitis*. Of course, many practitioners were using these agents for lowering cholesterol levels "off label." Probucol appears to have been "grandfathered in" and approved for lowering cholesterol levels even before the FDA formally recognized cholesterol lowering as a way of preventing coronary heart disease. In any case the CPPT results signaled a sea change in the position of the FDA on cholesterol-lowering drugs. Some in the agency continued to feel that demonstration of efficacy against clinical coronary heart disease should be required, but rather soon the agency began to approve cholesterol-lowering drugs without any formal commitment to a post-marketing Phase IV trial.

Cholestyramine (Questran™) and the closely related bile acid-binding resin, colestipol (Colestid™), originally available only as dry powders, have now been reformulated to tablet form and continue to be used effectively, especially when more than one agent is needed to reach therapeutic goals, as is true in many patients with inherited dyslipidemias. Another nonabsorbable bile acid sequestrant, colesevelam (WelChol™), has recently been added to the armamentarium and it appears to have fewer side effects (35).

CLOFIBRATE

The discovery of clofibrate is a curious example of serendipity in science. In the early 1960s, J.M. Thorp and colleagues, working at Imperial Chemical Industries Limited in England, were experimenting with the effects of steroid hormones on metabolism. It had been shown by Hellman *et al.* that the adrenal steroid androsterone could lower blood cholesterol levels when given intramuscularly but was without effect when given orally (36). Androsterone itself is not well absorbed from the intestine so Thorp and colleagues were hoping to find some way to increase its absorption. They found that clofibrate (α-*p*-chlorophenoxyisobutyrate), when administered orally along with androsterone, yielded a consistent fall in blood cholesterol levels (37;38). They inferred, logically enough, that the clofibrate was enhancing the absorption of androsterone, the presumed active component. Michael F. Oliver, at the University of Edinburgh, demonstrated the effectiveness of this combination in humans. He also reported that there was no response to the clofibrate when given alone at the same dosage. (It should be noted that Oliver only studied six patients at the time.) He, like Thorp, concluded that the effectiveness of the combination was attributable to the androsterone. In any case, after some small-scale safety studies, the combination of androsterone and clofibrate was put on the market under the trade name Atromid (each capsule containing 5.5 mg androsterone "dissolved in" 245 mg clofibrate). It was recommended "for the control of blood abnormalities in coronary heart disease." Concurrently the drug company started a large, multicenter trial to further evaluate safety and efficacy. In fact, plans were already afoot for a full-scale clinical trial of Atromid to see if lowering blood cholesterol would reduce risk of coronary events. Then something most unexpected happened. In 1963, a group of investigators at the Sloan-Kettering Institute in New York led by Dr. Leon Hellman reported that in their hands the clofibrate contained in Atromid, *without* the androsterone, was fully as effective as the combination (39;40). Subsequent studies by Oliver (41) and by other investigators quickly confirmed that the androsterone in the preparation could indeed be totally omitted with no change in efficacy, i.e. clofibrate alone was just as effective as the combination. The drug company did not lose a beat. They simply rechristened it "Atromid-S" instead of "Atromid" the "S" presumably denoting the Latin "sine" – without. Without androsterone, that is.

So, the theory – that clofibrate enhanced the oral availability of steroid hormones – was wrong, but clofibrate, by itself, was actually quite effective. It reduced blood cholesterol levels by about 15 percent and it did not appear to have any serious side effects. Two groups in the United Kingdom, one in Scotland under the leadership of Michael F. Oliver and one in England under the leadership of H.A. Dewar, went to work on designing randomized, placebo-controlled, double-blind studies of the effectiveness of clofibrate in preventing progression of coronary heart disease in high-risk individuals, men and women, who already had clinically evident heart disease (42;43). At about the same time, the World Health Organization sponsored an even larger cooperative primary prevention study, involving Edinburgh, Prague, and Budapest, to determine whether clofibrate treatment of men at high risk because of hypercholesterolemia but *without* clinically evident heart disease would reduce risk.

CLOFIBRATE REDUCES RISK OF INFARCTION

The results in the two UK studies (in patients who had had their first infarct or who had angina) were not entirely concordant but they did agree in general (44). Both showed that in patients who were having angina at the beginning of the study there was a significant reduction in mortality, especially in sudden deaths, and a significant reduction in events. In both trials there was also a reduction in the incidence of nonfatal infarcts, and again this was more marked in patients with angina. However, the degree of protection did not parallel the plasma lipid responses, and the authors speculated that clofibrate might have additional modes of action. The bottom line was that there was a trend toward protection against a second cardiac event. The next question was whether starting treatment earlier – patients who had not yet developed symptomatic disease – would be effective. That was the question addressed by a large-scale WHO-sponsored study.

The WHO study was started in 1965 and included over 15,000 healthy men, aged 30–59 (3). About 10,000 of these were men with high blood cholesterol levels, levels found in the upper third of the population. Half of these (5,000) were assigned to the placebo group and half to the clofibrate group. (A second, untreated control group was made up of men having blood cholesterol levels in the lower third of the population to allow comparison of their prognosis with that of the men treated with clofibrate.) Clofibrate, instead of giving the predicted

15 percent drop in blood cholesterol, only decreased it by an average of 9 percent. Despite that, over the approximately 5 years of the study, clofibrate reduced the overall incidence of nonfatal heart attacks by 20 percent, and this result was statistically significant at the canonical $p < 0.05$ level. Moreover, the degree of protection was greatest in the men with the highest initial levels of plasma cholesterol and in the men showing the greatest drop in cholesterol level in response to clofibrate. In other words, the data suggested that the preventive effect was directly ascribable to the cholesterol-lowering effect of the drug. (Interestingly these data conform nicely to the generalization reached many years later that whatever the nature of the cholesterol-lowering intervention – diet or drugs – a 1 percent drop in cholesterol level causes an approximate 2 percent reduction in cardiac risk.) If there were no catch, these findings lent strong support to the lipid hypothesis. But there was a catch.

CLOFIBRATE IS TOXIC

The catch was that total mortality, i.e. all deaths irrespective of cause, was actually *greater* in the clofibrate-treated group. The overall death rate was 20 percent higher and this difference was statistically significant ($p < 0.05$). So, even though clofibrate had reduced the number of nonfatal heart attacks significantly, overall it had apparently done more harm than good. To paraphrase an old aphorism: "The experiment was a success but the patient died."

The investigators looked carefully at the causes of death by category. There were no differences in the number of deaths from other vascular diseases nor from accidents or violence. The total number of cancers in the clofibrate group was slightly higher, irrespective of the site of the cancer, but that difference was not statistically significant. The excess deaths were predominantly associated with diseases of the liver, gall bladder, and intestine, including cancers in these locations. The authors hypothesized that clofibrate might be responsible by increasing cholesterol mobilization and excretion into the bile, but there was no direct evidence for that.

The key question was whether the increase in non-CHD deaths was due, directly or indirectly, to the lowering of serum cholesterol brought about by the drug or rather to some other pharmacologic effect of clofibrate not necessarily related to its cholesterol-lowering effect. If the former, then any other cholesterol-lowering regimen would be similarly toxic and research in the field might as well come to a complete halt. If the latter were true, then other pharmacologic or dietary approaches might very well reduce serum cholesterol and risk of heart attack without increasing deaths from other diseases. At the time no one could answer the question unambiguously and therefore an understandable conservatism held sway for some time regarding treatment of hypercholesterolemia.

Michael F. Oliver, who was a prime mover in the design and execution of these trials, and the many clinician-scientists who participated must have been bitterly disappointed in the outcome. Dr. Oliver, his close collaborator George S. Boyd, and their colleagues in Edinburgh had for many years done pioneering work on lipoproteins and atherosclerosis, especially on the role of hormones in the process (45;46). They must have felt reasonably sure there was a causal relationship between blood cholesterol and atherosclerosis or else they would not have invested so much time and energy in the design and execution of these clinical trials. It would be readily understandable that Oliver and the others involved would henceforth take a skeptical if not frankly negativistic view of any proposals to treat hypercholesterolemia. Oliver spoke out passionately against such recommendations for years, even after the 1984 Consensus Conference at which the National Heart, Lung and Blood Institute endorsed a national program for lowering blood cholesterol to prevent heart disease (47). He was indeed "The Cholesterol Pessimist" – possibly because the clofibrate trials had left scars.

Later studies, using more effective drugs for lowering cholesterol levels and involving many thousands of subjects, would clearly show that lowering cholesterol levels does not per se increase deaths from any other causes, including cancers, violence or accidents, or diseases of the gastrointestinal tract. The unfortunate side effects in the WHO study must have been due to some other effect of the clofibrate molecule or due to chance alone. Ironically, the fact that clofibrate treatment had in fact reduced the incidence of nonfatal myocardial infarctions by 20 percent seemed to be lost sight of. Later studies using the statins would show that not only coronary heart disease mortality but also total mortality is decreased when serum cholesterol is lowered, even when it is lowered by more than 40 percent. At the time, however, none of this was known, and the concerns about safety dampened efforts to make lowering cholesterol a high-priority public health goal.

It should be noted that an analogue of clofibrate, gemfibrozil, was introduced in the mid-1970s and shown to have comparable effects on serum lipid levels but without the side effects encountered in the clofibrate trials. Moreover, the Helsinki Heart Study, an intervention study in over 4,000 men at high risk because of dyslipidemia showed that gemfibrozil reduced incidence of coronary events by 34 percent *without affecting overall mortality* (48;49). Follow-up at 18 years showed a 32 percent lower relative risk in the original gemfibrozil-treated group and no difference in all-cause or cancer mortality (50).

PROBUCOL

The history of probucol is a peculiar one. Persuaded by the evidence supporting the lipid hypothesis, a number of companies were intensifying their search for

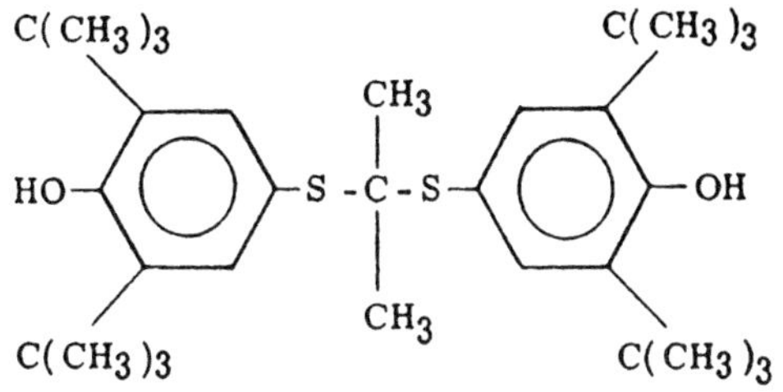

FIGURE 6.2   Structure of probucol. The trimethyl substituents flanking the ring hydroxyl groups make it a strong antioxidant.

nontoxic cholesterol-lowering drugs during the late 1960s and early 1970s. Dow Chemical Company was one of them. As a result of a random screen of their chemical library they came upon probucol (4,4'-(isopropylidenedithio)bis(2, 6-di-t-butylphenol)), see Figure 6.2.

The compound was effective in lowering cholesterol levels in mice, rats, and monkeys and it had no apparent toxic effects even at very high dosages (up to 5,280 mg/kg in rodents) (51). Preliminary tests quickly showed the compound to be highly effective in lowering cholesterol levels in humans as well (52). Plasma cholesterol levels were reduced by 10–20 percent. There was no reduction in triglyceride levels. The early studies of probucol have been extensively reviewed (53).

Probucol (Lorelco®) was approved by the FDA for cholesterol lowering even though at the time its mechanism of action was not known and even though neither it nor any other cholesterol-lowering drug had ever been tested in randomized trials for efficacy in preventing coronary heart disease. In a sense, probucol was grandfathered in. Subsequently the agency did not approve drugs specifically for lowering cholesterol levels until the results of the NIH Coronary Primary Prevention Trial became available. In any case, probucol was rather widely used clinically for many years despite the fact that it lowered HDL cholesterol even more than it raised LDL cholesterol (actually worsening the LDL/HDL ratio) and despite the fact that it had the potential to increase the QT interval in the electrocardiogram, potentially increasing the risk of clinically significant arrhythmias. There have been anecdotal reports of arrhythmias in patients treated with probucol (notably torsade de pointes) but these patients all had other possible reasons for their arrhythmias. In a post-marketing survey in Japan, QTc prolongation of more than 51 ms was seen in 2.9 percent of subjects with initially normal QTc values (<440 ms) and in none of those with initially prolonged intervals (>440 ms) (54). There have been no large-scale, placebo-controlled studies on this important issue of safety in long-term treatment. The PQRST study, discussed below, involved over 300 men divided into placebo and treatment groups and followed for 3 years, but no data were presented with regard to QT interval or arrhythmias.

At doses of 500 mg twice daily probucol reduced total cholesterol levels by 10–20 percent and side effects were mild and infrequent. Still practitioners hesitated to use a drug that consistently lowered HDL levels, an effect that might cancel out any gains from lowering LDL. Yet several reports showed, remarkably, that probucol was effective even in patients with homozygous familial hypercholesterolemia, a group notoriously resistant to treatment with diet or other drugs (55–57). The fall in total cholesterol in these homozygous patients (a total of 16 in these reports) was as great as or even greater than that seen in heterozygotes or in other dyslipidemic subjects. Moreover, after one or two years of treatment there was obvious regression of skin and tendon xanthomas, some actually disappearing. In the study of Yamamoto *et al.* the regression of xanthomas was documented quantitatively by measuring the width of the Achilles tendon as a function of time by xerography. These dramatic findings suggested that the drug was antiatherosclerotic *despite the drop in HDL.* They also suggested some antiatherosclerotic effect of probucol over and above its LDL-lowering potential.

A formal clinical trial designed to test whether probucol could inhibit progression of femoral atherosclerosis was undertaken in Sweden (58), under the tricky acronymic title PQRST (Probucol Quantitative Regression Swedish Trial). Over 300 men with severe femoral atherosclerosis were randomly assigned to placebo or probucol and quantitative femoral angiography was performed at baseline and at yearly intervals for 3 years. No beneficial effect of probucol could be seen. However, it should be noted that what was measured was the *total three-dimensional volume* of a defined femoral segment reconstructed from two views of the artery. Regression or progression of individual segments would have had only a small effect on the integrated total luminal volume, so small effects could have been overlooked. Also, all subjects were treated with large doses of cholestyramine, i.e. there was no control, placebo-only group. Finally, the lesions in the femoral artery were advanced, complex lesions. Nevertheless, here was a large-scale, controlled study of clinical effectiveness, and it was negative. Later studies would show, however, that probucol was indeed effective in reversing the growth of early carotid lesions (measurements of intima-media thickness by ultrasound) (59). In this study probucol, 500 mg/d, was just as effective as pravastatin, 10 mg/d.

It is unlikely that a large-scale trial of the effectiveness of probucol in preventing coronary events will ever be undertaken. The holders of the patent on probucol, Hoechst Marion Roussel, were told by the FDA that they would have to carry out a Phase IV clinical trial to demonstrate clinical efficacy. In view of the limited market share of probucol in the statin era and the pending expiration of their patent, they decided to withdraw probucol from the market rather than undertake a large, expensive clinical trial.

Recently there has been a renewed interest in probucol because it has been shown to be effective in preventing restenosis after balloon injury in animals (60) and after coronary angioplasty in humans (61).

ANOTHER BLOW: THE SOMEWHAT MESSY CORONARY DRUG PROJECT

By the early 1960s the lipid hypothesis was accepted by many specialists in the atherosclerosis research community but by no means all. Certainly it was not accepted by the average practitioner. The dietary intervention studies were strongly suggestive but not totally convincing. A "clincher" was needed; a study that would lower cholesterol levels more effectively and include enough subjects to give an unambiguous result. The National Heart and Lung Institute decided in 1960 to go all out, and the Coronary Drug Project got underway in 1965. It would finance a study using the best of the cholesterol-lowering drugs available, although their effectiveness was limited, and it would go for a truly large-scale study by enlisting the participation of many medical centers. A total of 8,341 subjects were randomized at 53 centers. The study would be done in men aged 30–64 who had already had a first myocardial infarction. Unfortunately, the project was launched rather hurriedly, and only after it was already underway did some rather serious problems begin to surface (16;62).

Four different cholesterol-lowering agents were selected for study: dextrothyroxine, estrogenic hormone, clofibrate, and nicotinic acid. Preliminary studies with D-thyroxine had demonstrated that, unlike the active natural hormone, it had no apparent effects on metabolism, blood pressure, and heart rate and yet retained the cholesterol-lowering effect. The operative word here was, unfortunately, *apparent*. Within 18 months of the start of the planned 5-year study an excess mortality was noted in patients who had had pre-existing arrhythmias. Over the next 2 years the mortality rate in the D-thyroxine-treated group as a whole kept rising, becoming almost statistically significant, and this arm of the study was discontinued. What had not been appreciated was that even a minimal level of increased thyroid hormone activity is enough to evoke arrhythmias and to have harmful effects on the cardiovascular system generally. D-thyroxine has *very little* hormonal activity – but *not none*.

The second drug chosen was estrogenic hormone, even though the plan was to study only men. The rationale here was that women before menopause have a much lower coronary heart disease risk than men of the same age and it was the conventional wisdom that it was the hormonal pattern in premenopausal women that protected them. Would a decreased risk of a second infarction be worth the inevitable feminizing effects? It seemed so to enough men to make up the study quorum. Sad to say, within a year and a half the rate of nonfatal heart attacks in the men on the high dose of estrogen (5.0 mg conjugated equine estrogen) was significantly *greater* than that in the controls (62) and it was discontinued. A few years later, the low dose regimen (2.5 mg) was also discontinued because of a small but statistically significant increase in all-cause mortality. There was also a suggestive increase in deep vein thrombosis and cancer. Needless to say, these effects of the estrogens in men were entirely unexpected at the time. The other two arms of the study were continued for the full 5-year follow-up (4). Coronary death rates on clofibrate were slightly less than in controls, but the difference was not statistically significant. There was no difference in total mortality. Worst of all, in the clofibrate group there was a significantly greater incidence of serious side effects, including angina pectoris, peripheral vascular disease, arrhythmias, and venous thrombosis. There was also a twofold increase in the incidence of gall-stones on clofibrate.

Finally, the good news! *The men on nicotinic acid showed a statistically significant decrease in nonfatal heart attacks.* While there was no decrease in overall mortality during the 5-year period of the study itself, when the men were evaluated about 9 years later the nicotinic acid-treated group showed an 11 percent lower overall mortality than the controls and this was statistically highly significant ($p < 0.0004$). It should be noted that among the drug treatments tried, nicotinic acid yielded the greatest drop in serum cholesterol, but this was still only about 10 percent (17).

This is an instructive illustration of the arbitrariness of our definitions of "significant." During the initial follow-up there was at best a tendency toward decreased overall mortality, and the conclusion reached was that nicotinic acid treatment does not decrease all-cause mortality. After the end of the trial the men went to their private physicians for management, so there was no systematic difference between them with respect to treatment. What was different? First, 11 years later the absolute numbers of total deaths was larger and, second, there may well have been a persisting benefit after the discontinuance of therapy that made fatal events less likely. Notably this was the first trial in which total mortality as well as coronary mortality was significantly affected. The finding cheered the "convinced" but did little to move the "unconvinced."

THE CHOLESTEROL CONTROVERSY AT ITS HEIGHT

As of 1983, there were two camps on the issue of whether the cholesterol–coronary heart disease data warranted recommending at least diet modification to lower elevated blood cholesterol. Perhaps the most vocal opposition came from Great Britain. Michael F. Oliver was on record to the effect that "reduction of raised serum cholesterol is a card of uncertain quality in the primary prevention of [coronary heart disease]" and that "reduction of raised serum cholesterol could lead to adverse biological changes" (63). He also advised, "It is probably of little value to reduce raised serum cholesterol concentrations in patients with overt [coronary heart disease]" (64). In contrast, the American Heart Association as early as 1961 had accepted the causal relationship between blood cholesterol and atherosclerosis and in that year recommended that people at high risk because of high blood cholesterol or strong family history of coronary heart disease be advised to modify their diet to avert heart attacks. In 1964, the Heart Association extended these dietary recommendations to the general public and in 1965 the Food and Nutrition Council of the American Medical Association made similar recommendations. These two associations represented American cardiology and American internal medicine and they felt the evidence was adequate – not iron clad, but adequate – to justify taking action.

At the same time, however, E.H. Ahrens, Jr, whose own pioneering clinical research showed conclusively that blood cholesterol could indeed be reduced by appropriate changes in diet, took exception to proposals to change the diet of the American public. In 1979 he wrote that such recommendations would be "unwise, impractical, and unlikely to lead to a reduced incidence of arteriosclerotic disease" (65).

I.D. Frantz, Jr and Richard B. Moore summarized the situation very aptly in a 1969 review (66) cited in Chapter 1:

> Few controversies have divided the medical community so sharply for such a long time as has the sterol hypothesis. The separation between the two points of view has become so extreme that, on the one hand, there are respected scientists who believe that the evidence is already so convincing that further clinical testing is unnecessary, financially wasteful and actually unethical; and, aligned against them, are equally respected scientists who believe that the total weight of evidence accumulated over the many years is too slight to justify further work along these lines.

Sad to say, this was still a reasonably good description of the situation 15 years later, in 1983, despite an ever-increasing number of epidemiologic studies, experimental animal studies, and additional dietary studies suggesting a causal relationship between blood cholesterol and coronary heart disease. We still needed a "you-can't-argue-with-this" type of study.

REFERENCES

1. Kinsell, L.W., Partridge, J., Boling, L., Margen, S., and Michael, G. 1952. Dietary modification of serum cholesterol and phospholipid levels. *J Clin Endocrinol Metab* 12:909–913.
2. Ahrens, E.H., Jr., Blankenhorn, D.H., and Tsaltas, T.T. 1954. Effect on human serum lipids of substituting plant for animal fat in diet. *Proc Soc Exp Biol Med* 86:872–878.
3. 1978. A co-operative trial in the primary prevention of ischaemic heart disease using clofibrate. Report from the Committee of Principal Investigators. *Br Heart J* 40:1069–1118.

4. 1975. Clofibrate and niacin in coronary heart disease. *JAMA* 231:360–381.

5. Altschul, R., Hoffer, A., and Stephen, J.D. 1955. Influence of nicotinic acid on serum cholesterol in man. *Arch Biochem* 54:558–559.

6. Altschul, R. 1950. *Selected Studies on Arteriosclerosis.* Charles C. Thomas, Springfield, Illinois.

7. Altschul, R. 1953. Inhibition of experimental cholesterol arteriosclerosis by ultraviolet irradiation. *N Engl J Med* 249:96–99.

8. Altschul, R. and Herman, I.H. 1954. Influence of oxygen inhalation on cholesterol metabolism. *Arch Biochem Biophys* 51:308–309.

9. Altschul, R. and Hoffer, A. 1958. Effects of salts of nicotinic acid on serum cholesterol. *Br Med J* 46:713–714.

10. Achor, R.W., Berge, K.G., Barker, N.W., and McKenzie, B.F. 1958. Treatment of hypercholesterolemia with nicotinic acid. *Circulation* 17:497–504.

11. Parsons, W.B., Jr. and Flinn, J.H. 1957. Reduction in elevated blood cholesterol levels by large doses of nicotinic acid; preliminary report. *J Am Med Assoc* 165:234–238.

12. Parsons, W.B., Jr. and Flinn, J.H. 1959. Reduction of serum cholesterol levels and beta-lipoprotein cholesterol levels by nicotinic acid. *AMA Arch Intern Med* 103:783–790.

13. Stamler, J. 1977. The coronary drug project – findings with regard to estrogen, dextrothyroxine, clofibrate and niacin. *Adv Exp Med Biol* 82:52–75.

14. Carlson, L.A. and Oro, L. 1962. The effect of nicotinic acid on the plasma free fatty acid; demonstration of a metabolic type of sympathicolysis. *Acta Med Scand* 172:641–645.

15. Carlson, L.A. 1963. Studies on the effect of nicotinic acid on catecholamine stimulated lipolysis in adipose tissue in vitro. *Acta Med Scand* 173:719–722.

16. 1972. The coronary drug project. Findings leading to further modifications of its protocol with respect to dextrothyroxine. The coronary drug project research group. *JAMA* 220:996–1008.

17. Canner, P.L., Berge, K.G., Wenger, N.K., Stamler, J., Friedman, L., Prineas, R.J., and Friedewald, W. 1986. Fifteen year mortality in Coronary Drug Project patients: long-term benefit with niacin. *J Am Coll Cardiol* 8:1245–1255.

18. Tunaru, S., Kero, J., Schaub, A., Wufka, C., Blaukat, A., Pfeffer, K., and Offermanns, S. 2003. PUMA-G and HM74 are receptors for nicotinic acid and mediate its anti-lipolytic effect. *Nat Med* 9:352–355.

19. Benyo, Z., Gille, A., Kero, J., Csiky, M., Suchankova, M.C., Nusing, R.M., Moers, A., Pfeffer, K., and Offermanns, S. 2005. GPR109A (PUMA-G/HM74A) mediates nicotinic acid-induced flushing. *J Clin Invest* 115:3634–3640.

20. Pike, N.B. 2005. Flushing out the role of GPR109A (HM74A) in the clinical efficacy of nicotinic acid. *J Clin Invest* 115:3400–3403.

21. McCormack, P.L. and Keating, G.M. 2005. Prolonged-release nicotinic acid: a review of its use in the treatment of dyslipidaemia. *Drugs* 65:2719–2740.

22. Carlson, L.A. 2005. Nicotinic acid: the broad-spectrum lipid drug. A 50th anniversary review. *J Intern Med* 258:94–114.

23. Siperstein, M.D., Nichols, C.W., Jr., and Chaikoff, I.L. 1953. Effects of ferric chloride and bile on plasma cholesterol and atherosclerosis in the cholesterol-fed bird. *Science* 117:386–389.

24. Siperstein, M.D., Chaikoff, I.L., and Reinhardt, W.O. 1952. C14-cholesterol. V. Obligatory function of bile in intestinal absorption of cholesterol. *J Biol Chem* 198:111–114.

25. Peters, J.P. and Vanslyke, D.D. 1946. *Quantitative Clinical Chemistry.* Williams and Wilkins, Baltimore.

26. Siperstein, M.D., Nichols, C.W., Jr., and Chaikoff, I.L. 1953. Prevention of plasma cholesterol elevation and atheromatosis in the cholesterol-fed bird by the administration of dihydrocholesterol. *Circulation* 7:37–41.

27. Bergen, S.S., Jr., Van Itallie, T.B., Tennent, D.M., and Sebrell, W.H. 1959. Effect of an anion exchange resin on serum cholesterol in man. *Proc Soc Exp Biol Med* 102:676–679.

28. Tennent, D.M., Siegel, H., Zanetti, M.E., Kuron, G.W., Ott, W.H., and Wolf, F.J. 1960. Plasma cholesterol lowering action of bile acid binding polymers in experimental animals. *J Lipid Res* 1:469–473.

29. Tennent, D.M., Kuron, G.W., Zanetti, M.E., and Ott, W.H. 1961. Plasma cholesterol concentrations in cockerels and dogs treated with bile acid binding polymer and cholesterol synthesis inhibitors. *Proc Soc Exp Biol Med* 108:214–216.

30. Hunninghake, D.B. 1991. Bile acid sequestrants. In *Drug Treatment of Hyperlipidemia*, B.M. Rifkind, ed. Marcel Dekker, Inc., New York:89–102.

31. 1984. The Lipid Research Clinics Coronary Primary Prevention Trial results. I. Reduction in incidence of coronary heart disease. *JAMA* 251:351–364.

32. 1984. The Lipid Research Clinics Coronary Primary Prevention Trial results. II. The relationship of reduction in incidence of coronary heart disease to cholesterol lowering. *JAMA* 251:365–374.

33. Van Itallie, T.B., Hashim, S.A., Crampton, R.S., and Tennent, D.M. 1961. The treatment of pruritus and hypercholesteremia of primary biliary cirrhosis with cholestyramine. *N Engl J Med* 265:469–474.

34. 1985. Addition to labeling of cholestyramine. *FDA Drug Bull* 15:7–8.

35. Davidson, M.H., Dicklin, M.R., Maki, K.C., and Kleinpell, R.M. 2000. Colesevelam hydrochloride: a non-absorbed, polymeric cholesterol-lowering agent. *Expert Opin Investig Drugs* 9:2663–2671.

36. Hellman, L., Bradlow, H.L., Zumoff, B., Fukushima, D.K., and Gallagher, T.F. 1959. Thyroid–androgen interrelations and the hypocholesteremic effect of androsterone. *J Clin Endocrinol Metab* 19:936–948.

37. Thorp, J.M. and Waring, W.S. 1962. Modification of metabolism and distribution of lipids by ethyl chlorophenoxyisobutyrate. *Nature* 194:948–949.

38. Thorp, J.M. 1962. Experimental evaluation of an orally active combination of androsterone with ethyl chlorophenoxyisobutyrate. *Lancet* 279:1323–1326.

39. Hellman, L., Zumoff, B., Kessler, G., Kara, E., Rubin, I.L., and Rosenfeld, R.S. 1963. Reduction of cholesterol and lipids in man by ethyl p-chlorophenoxyisobutyrate. *Ann Intern Med* 59:477–494.

40. Hellman, L., Zumoff, B., Kessler, G., Kara, E., Rubin, I.L., and Rosenfeld, R.S. 1963. Reduction of serum cholesterol and lipids by ethyl chlorophenoxyisobutyrate. *J Atheroscler Res* 3:454–466.

41. Oliver, M.F. 1962. Reduction of serum-lipid and uric-acid levels by an orally active androsterone. *Lancet* 1:1321–1323.

42. 1971. Trial of clofibrate in the treatment of ischaemic heart disease. Five-year study by a group of physicians of the Newcastle upon Tyne region. *Br Med J* 4:767–775.

43. 1971. Ischaemic heart disease: a secondary prevention trial using clofibrate. Report by a research committee of the Scottish Society of Physicians. *Br Med J* 4:775–784.

44. Dewar, H.A. and Oliver, M.F. 1971. Secondary prevention trials using clofibrate: a joint commentary on the Newcastle and Scottish trials. *Br Med J* 4:784–786.

45. Boyd, G.S. and Oliver, M.F. 1956. Endocrine aspects of coronary sclerosis. *Lancet* 271:1273–1276.

46. Oliver, M.F. and Boyd, G.S. 1954. The effect of estrogens on the plasma lipids in coronary artery disease. *Am Heart J* 47:348–359.

47. Oliver, M.F. 1985. Consensus or nonsensus conferences on coronary heart disease. *Lancet* 1:1087–1089.

48. Frick, M.H., Elo, O., Haapa, K., Heinonen, O.P., Heinsalmi, P., Helo, P., Huttunen, J.K., Kaitaniemi, P., Koskinen, P., Manninen, V., Maenpaa, H., Malkonen, M., Manttari, M., Norola, S., Pasternack, A., Pikkaraineen, J., Romo, M., Sjoblem, T. and Nikkila, E.A. 1987. Helsinki Heart Study: primary-prevention trial with gemfibrozil in middle-aged men with dyslipidemia. Safety of treatment, changes in risk factors, and incidence of coronary heart disease. *N Engl J Med* 317:1237–1245.

49. Huttunen, J.K., Heinonen, O.P., Manninen, V., Koskinen, P., Hakulinen, T., Teppo, L., Manttari, M., and Frick, M.H. 1994. The Helsinki Heart Study: an 8.5-year safety and mortality follow-up. *J Intern Med* 235:31–39.

50. Tenkanen, L., Manttari, M., Kovanen, P.T., Virkkunen, H., and Manninen, V. 2006. Gemfibrozil in the treatment of dyslipidemia: an 18-year mortality follow-up of the Helsinki Heart Study. *Arch Intern Med* 166:743–748.

51. Barnhart, J.W., Sefranka, J.A., and McIntosh, D.D. 1970. Hypocholesterolemic effect of 4,4'-(isopropylidenedithio)-bis(2,6-di-t-butylphenol) (probucol). *Am J Clin Nutr* 23:1229–1233.

52. Drake, J.W., Bradford, R.H., McDearmon, M., and Furman, R.H. 1969. The effect of [4,4'-(isopropylidenethio)bis(2,6-di-t-butylphenol)] (DH-581) on serum lipids and lipoproteins in human subjects. *Metabolism* 18:916–925.

53. Buckley, M.M., Goa, K.L., Price, A.H., and Brogden, R.N. 1989. Probucol. A reappraisal of its pharmacological properties and therapeutic use in hypercholesterolaemia. *Drugs* 37:761–800.

54. Yoshioka, K., Kosasayama, A., Yoshida, M., Toshikura, M., Nasu, K., Funabashi, S., Yamada, A., and Okada, S. 1999. Post marketing surveillance of 13 products: safety and effectiveness. *Pharmacoepidemiol Drug Saf* 8:31–43.

55. Rapado, A., Castrillo, J.M., and Diaz-Curiel, M. 1979. Use of diet and probucol. *Arch Intern Med* 139:1062–1063.

56. Baker, S.G., Joffe, B.I., Mendelsohn, D., and Seftel, H.C. 1982. Treatment of homozygous familial hypercholesterolaemia with probucol. *S Afr Med J* 62:7–11.

57. Yamamoto, A., Matsuzawa, Y., Yokoyama, S., Funahashi, T., Yamamura, T., and Kishino, B. 1986. Effects of probucol on xanthomata regression in familial hypercholesterolemia. *Am J Cardiol* 57:29H–35H.

58. Walldius, G., Erikson, U., Olsson, A.G., Bergstrand, L., Hadell, K., Johansson, J., Kaijser, L., and Lassvik, C. 1994. The effect of probucol on femoral atherosclerosis: the probucol quantitative regression Swedish trial (PQRST). *Am J Card* 74:875–883.

59. Sawayama, Y., Shimizu, C., Maeda, N., Tatsukawa, M., Kinukawa, N., Koyanagi, S., Kashiwagi, S., and Hayashi, J. 2002. Effects of probucol and pravastatin on common carotid atherosclerosis in patients with asymptomatic hypercholesterolemia. Fukuoka Atherosclerosis Trial (FAST). *J Am Coll Cardiol* 39:610–616.

60. Ferns, G.A., Forster, L., Stewart-Lee, A., Konneh, M., Nourooz-Zadeh, J., and Anggard, E.E. 1992. Probucol inhibits neointimal thickening and macrophage accumulation after balloon injury in the cholesterol-fed rabbit. *Proc Natl Acad Sci U.S.A.* 89:11312–11316.

61. Tardif, J.C., Cote, G., Lesperance, J., Bourassa, M., Lambert, J., Doucet, S., Bilodeau, L., Nattel, S., and de Guise, P. 1997. Probucol and multivitamins in the prevention of restenosis after coronary angioplasty. Multivitamins and Probucol Study Group. *N Engl J Med* 337:365–372.

62. 1970. The Coronary Drug Project. Initial findings leading to modifications of its research protocol. *JAMA* 214:1303–1313.

63. Oliver, M.F. 1981. Serum–cholesterol the knave of hearts and the joker. *Lancet* 2:1090–1095.

64. Oliver, M.F. 1981. Lipid lowering and ischaemic heart disease. *Acta Med Scand* 651 Suppl:285–293.

65. Ahrens, E.H. 1979. Dietary fats and coronary heart disease: unfinished business. *Lancet* 2:1345–1348.

66. Frantz, I.D., Jr. and Moore, R.B. 1969. The sterol hypothesis in atherogenesis. *Am J Med* 46:684–690.

CHAPTER 7

The 1984 Coronary Primary Prevention Trial

The Keystone in the Arch of Evidence Linking Blood Cholesterol to Coronary Heart Disease

As of the early 1980s, despite the wealth of evidence from experimental animal models, the extensive epidemiologic evidence, the powerful genetic evidence, and the strongly suggestive clinical intervention trial results, most clinicians still remained unpersuaded regarding the relevance of the lipid hypothesis. What was needed was a well-designed, large-scale, long-term, double-blind study demonstrating a statistically significant impact of treatment on coronary heart disease events. The National Institutes of Health (NIH) had laid the groundwork for such a study as early as 1970 but the study was not completed and the results were not available until 1984. This study, the Coronary Primary Prevention Trial (CPPT) (1;2), showed that treatment with a bile acid-binding resin reduced major coronary events in hypercholesterolemic men by 19 percent ($p = 0.05$). The NIH followed this up with a national Consensus Development Conference on Lowering Blood Cholesterol to Prevent Heart Disease. For the first time the NIH now went on record to advocate screening for hypercholesterolemia and urging aggressive treatment for those at high risk. The Institute initiated a national cooperative program to that end, the National Cholesterol Education Program. For the first time, preventing coronary heart disease became a national public health goal.

THE NATIONAL INSTITUTES OF HEALTH GOES INTO HIGH GEAR

BIRTH OF THE LIPID RESEARCH CLINICS

As early as 1970 many of the country's experts in atherosclerosis and preventive cardiology – and the American Heart Association – were already convinced

The Cholesterol Wars by D. Steinberg.
Copyright © 2007 Elsevier Inc. All rights of reproduction in any form reserved.

that there was a causal connection between blood cholesterol and coronary heart disease. However, no one could be certain how firm that connection was nor how much impact treatment to lower cholesterol levels would have. As a result almost no nonspecialists – and in fact very few practicing internists or cardiologists – were paying very much attention to their patients' high blood cholesterol levels in 1970; coronary heart disease continued to be the Number One cause of death. What to do?

The National Institutes of Health realized that the launching of a national program to treat high blood cholesterol levels would be enormously complex and expensive. They could not justify that expense without first having ironclad proof that treatment would work. In any case, the medical community would have to be convinced before they could be expected to make serious efforts to implement any proposed treatment programs. Even though the *cumulative* evidence was impressive, the direct clinical intervention trials were individually weak. What was missing was an airtight study – a "clincher" – that would bring the skeptics into the fold and mobilize the medical community.

In June 1970, Theodore Cooper, Director of the National Heart and Lung Institute, asked Donald S. Fredrickson, then Director for Intramural Research, to convene an expert panel, the Panel on Hyperlipidemia and Atherosclerosis, to advise him on how the Institute should proceed with respect to prevention of heart disease related to hypercholesterolemia. At the time the Panel was convened, Fredrickson and his colleagues, Robert I. Levy and Robert S. Lees, had just published a highly influential series of reviews in the *New England Journal of Medicine* (3–5). They had proposed a classification of abnormalities of serum lipids based on total cholesterol and triglyceride levels plus the lipoprotein pattern revealed by paper chromatography. Despite the pioneering lipoprotein work of Gofman (6;7), the concept of lipoproteins and the classification of patients based on these lipoprotein patterns was still foreign to most practitioners; also there were technical problems associated with lipoprotein measurements still to be resolved. One question the Panel was asked to address was whether the Institute should establish a network of lipid centers of excellence across the country that could standardize methods of lipid and lipoprotein analysis. These centers, to be designated "Lipid Research Clinics," would consult with local hospitals and clinics and help them train their own staffs in the use of these new methods. The Panel unanimously endorsed this proposal.[1]

Equally important, or possibly even more important, the Panel was also asked: "Do you believe the evidence is sufficient to warrant detection of and some form of individual treatment of hyperlipidemia?" Of the 21 experts, 20 answered yes. However, they recognized that the evidence that such treatment would reduce heart attack rates and by how much was still limited. So, they went on to recommend that the program must include a randomized intervention trial to determine the effect of treatment of hypercholesterolemia on atherosclerotic complications.

As a member of that Panel, I remember that some, including Fredrickson himself, felt strongly that the establishment of the network of Lipid Research Clinics should not wait until the nature of the randomized trial was agreed upon. Fredrickson was concerned that the intervention trial would be so costly that it might drain away funds from the other parts of the program. Others, including Levy, felt strongly that a definitive intervention trial should be an integral part of the package and get a very high priority. As it turned out, fortunately, the Lipid Research Clinics program was approved and funded with almost no opposition. A new branch of the Heart and Lung Institute, the Lipid Research Branch, was established to oversee the implementation of these recommendations and Levy was given the responsibility of heading it. Twelve university centers were successful in the competition for Clinic grants and these were up and running within a year or so.[2]

Planning for the randomized trial (the Coronary Primary Prevention Trial (CPPT)) was deferred but, fortunately, for only about a year. It was then proposed that each of the Lipid Research Clinics would participate in a multicenter trial to test definitively the lipid hypothesis.

DESIGNING THE TRIAL

The decision to launch such a complex study was not easy. It would be painfully expensive and it would take an enormous amount of planning and years of detailed implementation but it had to be done. Fredrickson voiced his own ambivalence in a witty 1968 article entitled "The field trial: some thoughts on the indispensable ordeal" (8). He began by suggesting that the first ever field trial was actually carried out in the Garden of Eden. Like so many subsequent field trials, he went on, it was roundly criticized because:

1. the experimental protocol had received inadequate prior consideration;
2. the population sample was too small; and
3. the study consumed too large a fraction of the then available gross national product.

Despite his tongue-in-cheek misgivings, he committed the National Heart and Lung Institute to a program that included a clinical trial as a major component.

In 1971, I was asked, along with John W. Farquhar from Stanford University, to co-chair an NIH Committee[3] that would design the protocol for the CPPT. Our Committee ran into many knotty problems, some theoretical, some ethical, and some just very difficult pragmatic problems. The discussions of the Committee spanned almost 2 years. The study would be named the Coronary

Primary Prevention Trial (CPPT) of the Lipid Research Clinics. The final result would not become available until 1984 – 13 years and about $150 million later.

The first thing the planning committee had to do was to decide on the mode of intervention. From one point of view, diet would have been the preferred treatment because then the issue of toxicity would not arise. However, diet was not really a viable choice. The degree of cholesterol lowering would be limited, and, unless huge numbers of subjects were studied (which would almost certainly make the costs unacceptable), the effects might wind up as marginal. Furthermore, a double-blind diet study would be all but impossible to design and fund. An NIH committee of consultants under the chairmanship of Edward H. Ahrens, Jr, had previously done an intensive, year-long study of the feasibility of a double-blind diet trial in the general population (9). They even considered an elaborate design in which fat-containing foods (meats, spreads, and dairy products) would be specially processed so that neither participants nor investigators could tell whether they contained saturated or polyunsaturated fats. The foods would carry bar coded labels and be issued to participants from a central warehouse. This would have enabled a double-blind study design. However, Ahrens' committee concluded that such a design while feasible in principle would be out of the question in practice; it would simply be forbiddingly expensive. The double-blind diet–heart design was a nonstarter.

What then were the alternatives? The only effective drugs in use at the time were clofibrate, nicotinic acid, and cholestyramine; none of these was ideal, as discussed below. In fact some researchers seriously doubted that the lipid hypothesis could ever be proved definitively using drug treatment. This pessimistic outlook led Henry Buchwald, a surgeon at the University of Minnesota, to explore the feasibility of a surgical approach (10). He showed that cholesterol levels could be lowered 20–25 percent using a modified intestinal bypass operation, and he initiated a long-term study to test whether this would decrease the risk of coronary heart disease. Obviously this intervention would hardly lend itself to a double-blind study. Also, asking volunteers to undergo major abdominal surgery with no absolute guarantee that there would be a benefit seemed daunting. Nevertheless, it was briefly considered. Parenthetically it should be noted that Buchwald stuck with his Program on the Surgical Correction of Hypercholesterolemia and eventually showed that treated subjects did have a significant decrease in coronary heart disease events and a decrease in total mortality compared to age-matched controls (11).

THE CANDIDATE DRUGS

Clofibrate was quite effective, lowering blood cholesterol by about 20 percent, and it had already been reported to reduce significantly cardiac end points in high-risk men (12). However, it was more effective in lowering VLDL than LDL and further-more its use had been associated with significant increases in gallstones (13) and other diseases of the gastrointestinal tract. Later studies would show that the drug, while having a favorable effect on cardiac events, actually increased overall mortal-ity (14). Our decision to pass up clofibrate was a fortunate one.

Nicotinic acid was effective but not an easy drug to take. In the formulations available at the time, it caused uncomfortable flushing and itching in a large fraction of patients. More seriously, it could impair liver function and, while such a side effect was relatively uncommon, our committee was unwilling to expose patients in the study to *any* unnecessary risks. Furthermore, the flush-ing would disclose which subjects were getting nicotinic acid and which placebo. So nicotinic acid might not be safe and the study could not be effec-tively double-blinded. Pass again.

Cholestyramine was effective. At full dosage (24 g/d) it reduced total blood cho-lesterol by 20–25 percent and LDL cholesterol by 30–35 percent (15). Because it was totally nonabsorbable it would predictably be free of systemic toxic side effects. However, it was at the time only available as a sandy powder that needed to be stirred in water or juice and gulped down. To be fully effective it had to be taken in doses of 24 g/d. That meant bravely downing two packets three times daily. Moreover, a significant percentage of patients taking large doses experienced bloating, constipation, or diarrhea due to local irritation of the intestinal wall.

So here was a drug that some patients would predictably find almost intoler-able, yet it was both safe and effective. Could we expect to persuade 3,800 men to take this gritty stuff regularly for 7 years? We elected to go with it, based mainly on its freedom from systemic toxicity. We would just have to grapple somehow with the problem of patient compliance and come up with imagina-tive ways to get the cholestyramine down and keep the morale up for 7 years.

Another tough problem was that of manufacturing a placebo that could not be distinguished from the active drug. Mead Johnson and Co., the makers of Questran, came through with polymer beads the same size and color as the active cholestyramine but with no ion-exchange groups on it. At the end of the trial the men were asked to say whether they had been in the treatment group or in the placebo group. Almost exactly 50 percent of the men got it right, as expected by chance alone. They couldn't tell the difference.

THE FINAL STUDY DESIGN

Our committee, after many meetings – with input from clinicians, statisticians, epidemiologists, lipid specialists – settled on a protocol after about 2 years, and a green light was given to start recruitment of patients. The study cohort would consist of about 3,800 men, aged 35–59, with no history of coronary heart disease and no signs of current disease. However, these men would be at high risk because of a total blood cholesterol level of 265 mg/dl or higher (men with cholesterol levels in the 95th percentile for this age group).

Recruiting 3,800 men fitting this description and willing to volunteer for a minimum of 5 years seemed straightforward but it proved to be a formidable undertaking. The original plan was that each Clinic would ask community physicians to refer patients who met the protocol requirements from their private practice. In addition, clinical laboratories and blood banks would be asked to identify (with permission) patients whose blood cholesterol was over 265. These approaches failed miserably. Practitioners simply weren't measuring cholesterol levels, and the number of blood donors was much lower than had been expected. The plan had optimistically called for completion of the recruitment phase in 18 months. Ten months into the study only 74 of the required 3,800 participants had been recruited and started on the protocol. First, men already seeing a physician probably already had coronary heart disease and were therefore not eligible for the study. Second, cholesterol levels were not yet routinely measured so there was no large existing database to draw on. There was a bit of a panic in the central office. Each year added to the recruitment period meant an additional year added to the length of the overall study – another $25 million or so. Radically different recruitment strategies were going to be needed.

It became clear that the CPPT would have to resort to mass public screening. It was going to be necessary to "go public" and do blood cholesterol levels in a random screening. The Lipid Research Clinic in St. Louis led the way. The Director, Gustave Schonfeld and the CPPT Director, Joseph L. Witztum, enlisted the help of a professional public relations firm to plan their campaign. This firm also handled public relations for McDonnel-Douglas, and the company agreed to let Witztum's recruiting team come in and draw blood samples from over 10,000 employees. The firm also provided entrée to some department stores they represented. Later, recruiting booths were set up at the Cardinals' games and the Rams' games. The "come on" was a free cholesterol measurement, and there were a goodly number of takers. Other Lipid Research Clinics adopted similar mass-screening strategies. Still, it took almost 3 years before the last subject was randomized. Eventually a total of almost 500,000 men nationwide had to be screened over the 3-year period from 1973 to 1976 before the full cohort of 3,800 participants was finally recruited.

POTENTIAL ETHICAL PROBLEMS ASSOCIATED WITH THE PLACEBO GROUP

All the men in this study had extremely high cholesterol levels and were, if the lipid hypothesis were correct, at a high risk of having a heart attack. This posed the sticky ethical question, common to all such intervention studies: would it be justifiable to leave the placebo group untreated? The answer today would be a definite "No," because today the lipid hypothesis has been proved. At the time, however, the lipid hypothesis was still just that – a hypothesis in the process of being tested. In fact, at the time, the volunteers for this project, even if they had consulted with their internist, would either have received no treatment at all or, at most, would have been given advice about diet. It was decided to have all participants follow a modest cholesterol-lowering diet such as their practitioners might have recommended. That diet was deliberately designed so that it would reduce total cholesterol by only about 5 percent. This would weaken the study by reducing the difference in cholesterol levels between the two groups and thus dilute the effect of the cholestyramine. However, based on the clinical experience available at the time, cholestyramine at full dose was expected to decrease cholesterol levels by about 30 percent. That would be more than enough to give a definitive result even though the diet reduced the cholesterol level somewhat in both groups. In any case, diet treatment was felt to be called for since some practitioners (though not many) were already recommending diet modification to patients with very high cholesterol levels.

THE TRIALS AND TRIBULATIONS OF THE CPPT TRIAL DIRECTORS

From the very beginning, the CPPT was a trial not only of the lipid hypothesis but also a trial for those running it. Many had had little previous experience with large-scale trials. Some, more interested in bench science, were actually a bit resentful of the time the trial would take away from their laboratory research programs. However, they were prepared to put their shoulders to the wheel because they recognized that the trial was of pivotal importance. The central Program Office at the NIH in Bethesda was headed initially by Robert I. Levy and then by Basil Rifkind. That central office played an absolutely essential role, taking the Trial Directors and staff members of the 12 collaborating centers by the hand and leading them through the thickets of clinical trial research. It was a complex operation. Each center employed physicians, nurses, dietitians, laboratory personnel, adherence counselors, and clerks, a total of about 30 full-time employees. The annual budget at each center was close to $2 million. Because many of the Trial Directors were in the program mainly out of a sense of obligation "to help get the job done," they sometimes lost patience with this sometimes monotonous year-after-year routine. The clinic staff worked tirelessly to improve compliance, but taking six packets of cholestyramine every day for more than 5 years was more than most of the men could manage, however good their intentions. Some brave souls managed it and in that group the final results were quite dramatic. But over-all compliance was disappointing. Yet, as discussed below, the study proved to be statistically significant but barely so.

Rifkind and the other administrators in the Bethesda Program Office had the responsibility to see that every "i" was dotted and every "t" crossed. After all, the outcome of this trial might determine whether or not the NIH would join the battle against cholesterol as a cause of myocardial infarction. Moreover, it was predictable that the results would be examined with a fine-tooth comb, espe-cially by the cholesterol skeptics. And it was going to cost about $150 million over all so it had better be of the highest quality. All these factors helped account for a certain amount of tension between the central Program Office and the indi-vidual Trial Directors out on the front line. The Directors chafed under the rigid insistence of the Program Office on strict adherence to protocol, the frequent joint meetings that meant travel sometimes far from home base, and a sometimes "no-discussion-allowed" approach. At the same time the Trial Directors recog-nized and applauded Rifkind for his outstanding job at the Program Office and, ultimately, for bringing the ship safely into port.

THE CPPT COMES IN WITH THE PROOF

When the CPPT trial ended, the 3,806 participants had been followed for an aver-age of 7.4 years. The degree of cholesterol lowering achieved with cholestyramine was, disappointingly, much less than had been expected: only a 13.4 percent reduction in total cholesterol and a 20.3 percent reduction in LDL cholesterol. Still, the number of events (definite coronary heart disease death and/or nonfatal heart attack) was 19 percent lower in the treated group with a p-value of <0.05, statistically significant although barely so (2), see Figure 7.1.

It was a narrow squeak. The CPPT came frighteningly close to joining the early dietary trials as "case not proved." In fact some criticized the investiga-tors for employing a one-tailed t-test because testing a treatment that might have serious adverse effects does require the use of a two-tailed t-test (16). However, cholestyramine was *not* being evaluated as a new drug; it was being used as an established means of lowering cholesterol levels to test the lipid hypothesis. So the use of the one-tailed statistic was appropriate. In any case, the results of the CPPT were considerably strengthened by the concordant and highly significant decreases in secondary end points: development of anginal pain decreased by 20 percent ($p < 0.01$), and development of a positive exer-cise electrocardiogram reduced by 25 percent ($p < 0.001$).

Another important point is that the published final results were reported, according to the standard practice for such studies, on *all* the men randomized without respect to whether they had or had not actually taken the prescribed six

Reprinted from the Journal of the American Medical Association
January 20, 1984 Volume 251
Copyright 1984, American Medical Association

Original Contributions

The Lipid Research Clinics Coronary Primary Prevention Trial Results

I. Reduction in Incidence of Coronary Heart Disease

Lipid Research Clinics Program

FIGURE 7.1   The 1984 paper presenting the results of the Lipid Research Clinics Coronary Primary Prevention Trial (1). This was the first truly large, double-blind, placebo-controlled trial showing unambiguously that lowering cholesterol levels in high-risk men (using cholestyramine) could significantly reduce risk of myocardial infarction. Source: (1).

packets of cholestyramine daily (24 g). In fact a large number of the men readily admitted that they simply could not handle the drug and stopped taking it altogether within weeks of the start of the trial. Others stayed with it but reduced the number of packets from the prescribed six daily to as low as two or three daily. At each clinic visit the men were given a large supply of packets, more than they would need to carry them until the next visit. One of the staff nurses would count the number of packets that had been used, the "packet count." During the first year the average daily "packet count," which should have been six, was only 4.2. By the end of the study it had dropped to less than four. This undoubtedly accounted for the discrepancy between the expected drop in cholesterol levels and the much more modest drop actually observed. When the event data for those men who had taken the full dose of six packets daily were analyzed separately, it was found that these men had had a 35 percent drop in total cholesterol and a 49 percent drop in event rate (Figure 7.2)!

For the study as a whole, the reduction in risk was proportional to the reduction in cholesterol level, as predicted by the lipid hypothesis. Despite the poor compliance and the smaller than expected drop in cholesterol levels, the study had made its point.

The relationship between the percentage lowering of cholesterol level and the percentage reduction in incidence of coronary events was consonant with the results in previous intervention trials (Figure 7.3).

This was despite the fact that the data compared included both drug and diet trials and both primary and secondary trials. The outlier "H" in Figure 7.3 represents

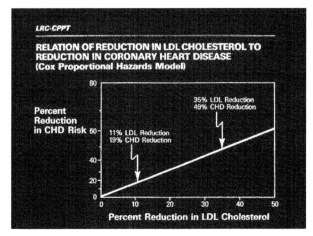

FIGURE 7.2   Percentage risk reduction versus percentage lowering of total serum cholesterol in the Coronary Primary Prevention Trial. The arrow to the left indicates the relationship in the cohort as a whole. However, as discussed in the text, many of the men stopped taking the drug almost from the first day and very few were able to take the full prescribed dose of 24 g/day. The arrow to the right indicates the relationship for those men who did manage to take the full dose. Source: data from references (1;2).

the data for the dextrothyroxine-treated group in the Coronary Drug Project (17), in which there was manifest cardiotoxicity including some fatal arrhythmias. The outlier "D" represents the extraordinary reduction in events achieved in the Newcastle clofibrate trial (18). The data suggested, but by no means proved, that no matter what the interventional modality the determinant of response was the degree to which cholesterol levels were decreased. (Most of the earlier studies did not include LDl cholesterol data.) Later trials, using the more potent statins would add points to the right of this graph and some of those points have been appended to Figure 7.3, viz: L, Scandinavian Simvastatin Survival Study (19); M, West of Scotland Coronary Prevention Study (20); N, CARE Study (21); O, British Heart Protection Study (22); and P, Program on the Surgical Control of Hypercholesterolemia (11). The degree of cholesterol lowering was greater in the later studies and the reduction in risk was greater but, at least to an approximation, the fit of the new data to the slope was rather close. A formal comparison of the pre-statin clinical trial results with the statin trial results suggests that the slope for the statin trials is slightly steeper but that the bulk of the benefit can be attributed to the drop in cholesterol levels rather than pleiotropic effects (23).

There was joy in 12 Mudvilles – in the 12 participating Lipid Research Clinics. At each clinic the men who had unselfishly volunteered and stayed the

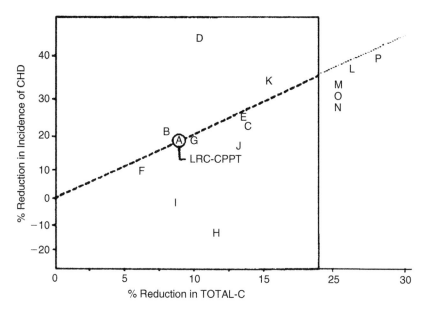

FIGURE 7.3   A modified version of Figure 3 from the CPPT report showing that the results in terms of percentage risk reduction versus percentage reduction in total cholesterol level were comparable to the results in previous studies, whether using drugs or diet to lower the cholesterol level. Points representing results from three of the early statin trials and the POSCH trial (ileal bypass to lower cholesterol level) have been added: L, Scandinavian Simvistatin Survival Study (19); M, West of Scotland study (20); N, CARE study (21); O, the British Heart Protection Study (22); and P, the Program on the Surgical Control of the Hyperlipidemias (POSCH) study (10;11).

course were invited together with their families to join the staff in celebration of the outcome. These men had persevered for 7 to 10 years in a very demanding regimen with a decidedly unattractive medication. Their contribution deserved and received appropriate recognition.

RECEPTION BY THE PROFESSION AND BY THE PRESS

The major medical journals around the world hailed the results of the trial as finally providing the rationale for treating hypercholesterolemia. The *Medical Journal of Australia* featured a lead article by Leon A. Simons entitled "The lipid hypothesis is proven" (24). He concluded with this: "The LRC-CPPT has given a new respectability and credibility to the dietary and pharmacologic management of hypercholesterolemia."

Postgraduate Medicine featured a Nutrition Highlights article by Richard N. Podell entitled "Coronary disease prevention. Proof of the anticholesterol pudding" (25). Even Michael F. Oliver, perhaps the most vocal skeptic over the years, wrote an editorial for the *British Medical Journal* entitled "Hypercholesterolaemia and coronary heart disease: an answer" (26). However, with characteristic pessimism, he warned that these CPPT results only applied to men with very high cholesterol values, that there was no guarantee that other drugs would be of benefit, and that the study did nothing really to settle the diet–heart problem.

Other researchers were also decidedly underwhelmed. George W. Mann, Associate Professor of Biochemistry at Vanderbilt University College of Medicine, had this to say about the CPPT directors: "They have held repeated press conferences bragging about this cataclysmic break-through which the study directors claim shows that lowering cholesterol lowers the frequency of coronary disease. They have manipulated the data to reach the wrong conclusions." And later: "The managers at NIH have used Madison Avenue hype to sell this failed trial in the way the media people sell an underarm deodorant" (27).

THE NATURE OF THE CRITICISMS

While the cholestyramine group in the CPPT showed significantly fewer fatal and nonfatal heart attacks, there was no decrease in total mortality. Was the treatment *increasing* mortality from some other diseases? What would be the point of preventing heart attacks if you just die of something else? This was perhaps the most troublesome criticism of the report and had to be taken seriously.

However, three considerations made it most unlikely that this was the case. First of all, there was not a *statistically significant* difference in total mortality. All-cause mortality was actually 7 percent *lower* in the treated group but that difference was not significant.

Second, the number of patients studied would have had to be considerably larger before one should have expected to see a statistically significant decrease in total mortality. As nicely pointed out by Collins *et al.* (28), the power of trials to detect a difference in total mortality depends very much on the numbers of subjects and has to be much larger than the numbers needed to detect a significant difference in coronary heart disease mortality, especially in primary prevention studies. The later, much larger statin studies *were* large enough and they *did show* a significant decrease in all-cause mortality (19;22). The point is that the apparently negative results with respect to all-cause mortality in the smaller, earlier trials with less effective cholesterol lowering do not allow a conclusion one way or the other regarding effects on all-cause mortality.

Finally, there was no statistically significant increase in deaths from any single disease or disease category. If cholestyramine itself or the lowering of blood cholesterol were toxic it might be expected to show up as an increased death rate in one or a few categories but that was not the case. Most of the noncoronary deaths were due to cancer and the numbers were almost identical in the two groups: 15 versus 16.

Much was made of the category of deaths in the CPPT that were lumped together (for reasons not entirely clear) as "traumatic deaths." These included accidents, homicide and suicide. Again, the difference was not statistically significant but the difference was large enough to be disturbing – only 4 in the placebo group but 11 in the cholestyramine group. A similar small (but nonsignificant) excess of deaths due to violence was reported in a study using gemfibrozil to lower cholesterol levels (29). Concern was expressed that *any* intervention that lowered blood cholesterol might affect cellular levels of cholesterol, especially in the brain, and in this way somehow induce aberrant behavior leading to violence (30;31). Wysowski and Gross (32) looked into the details of the deaths in these studies and made a persuasive case that there was really no evidence to support such a hypothesis. First of all the serum cholesterol levels in the accident victims were not particularly low, averaging 250 mg/dl at their last clinic visit. The one homicide death in the CPPT was actually not the murderer but rather the victim, surprised and shot by a burglar! Does taking cholestyramine or lowering your cholesterol level make you a more likely victim? Later studies with larger numbers of subjects and even more effective cholesterol-lowering drugs would establish definitively that lowering cholesterol levels actually reduces *both* coronary events and total mortality (19–22), but that was not yet firmly established in 1984. If lowering cholesterol levels was to become a national public health policy, literally millions of people might be treated. Even a small toxic effect might have drastic consequences. So these concerns had to be taken seriously. The CPPT studied only men aged 35–59, all with extremely high cholesterol levels. Many difficult decisions would have to be made about the extent to which the CPPT findings, taken together with the totality of evidence available from earlier studies, justified extrapolation to other subsets of the population.

THE 1984 NIH CONSENSUS DEVELOPMENT CONFERENCE ON LOWERING BLOOD CHOLESTEROL TO PREVENT CORONARY HEART DISEASE

With the results of the CPPT in hand, the NIH needed to decide what specific actions, if any, it should take. As one Australian lipid expert wrote (33), it was

"Time to treat cholesterol seriously." Even if it were accepted that an elevated blood cholesterol was an important causative factor, what levels called for treatment? Using which drugs or/and diets? A key question not directly answered by the LRC trial was whether lowering cholesterol levels to a comparable degree by *dietary* means rather than by cholestyramine treatment would give a comparable decrease in events. This seemed likely because the decrease in risk in this study was similar to that in the early studies using dietary treatment, but this could not be assumed. What were the potential hazards? At what age should treatment be started? In both men and women? The shape a national policy would take depended on the answers to these questions. Only if there were a true consensus would the NIH be prepared to establish new policy. The NIH had never before taken a position on how to deal with hypercholesterolemia, except in the rare, very severe genetically determined forms. The NIH had a well-established mechanism for getting such advice: the Consensus Development Conference.

Early in 1984 Basil Rifkind, Chief of the Lipid Metabolism-Atherogenesis Branch of the National Heart, Lung and Blood Institute, called and asked me if I would chair and help plan a Consensus Development Conference on "Lowering Blood Cholesterol to Prevent Heart Disease." As with other such NIH conferences, one of the Institutes proposes such a Conference, but the overall responsibility for planning it and appointing the panel of experts is vested in an independent NIH office, the Office of Medical Applications of Research (OMAR). That Office is independent of the individual Institutes, both in budget and in function. This independence is critical in keeping the Consensus Development Conferences as free of bias as possible. If the issue is important and if it seems there *might* be a chance for a consensus position, OMAR puts its machinery into operation to plan and organize the Conference. Rifkind's proposal had been accepted and they had agreed on my appointment to chair the Conference.

HOW THE NIH CONSENSUS CONFERENCES WORK

It may be in order to record some of the details of how this Conference was planned and by whom because the objectivity of the panelists and the correctness of their conclusions were publicly questioned by some (34–36).

Our Conference followed the pattern common to all such NIH Consensus Development Conferences: a Planning Committee[4] proposes the specific scientific questions to be addressed, lists the kinds of expertise that should be represented on the Consensus panel, and nominates a Chair. The final approval of all aspects of these proposals rests with the Office of Medical Applications of Research, which at the time was headed by Itzhak Jacoby. The Chair then works closely with the Planning Committee to refine the questions to be addressed and to select the experts invited to sit on the panel. The Committee also selects topics and speakers for a day and a half symposium in Bethesda, at which the scientific database relevant to the questions will be presented and discussed. Presentations at this symposium are made by experts not on the Consensus Panel itself. The symposium is widely publicized, and attendance is open to all interested parties: researchers, clinicians, other health professionals, food and drug manufacturers, lawyers, and the general public. Generally over 500 attendees come and participate in the discussion.

MEMBERS OF THE PANEL AND THE PANEL'S CHARGE

The specific questions to be addressed by our Panel were:

1. Is the relationship between blood cholesterol levels and coronary heart disease causal?
2. Will reduction of blood cholesterol levels help prevent coronary heart disease?
3. Under what circumstances and at what level of blood cholesterol should dietary or drug treatment be started?
4. Should an attempt be made to reduce the blood cholesterol levels of the general population?
5. What research directions should be pursued on the relationship between blood cholesterol and coronary heart disease?

Fourteen experts were invited to join the Panel and all accepted. They represented a wide span of disciplines, including biochemistry, endocrinology, cardiology, pathology, epidemiology and statistics, preventive medicine, and family medicine. Because our recommendations might have significant economic and legal implications, we also had two lawyers on the Panel, one who was a past-chairman of the American Heart Association, and one who was a public interest attorney.[5] Over the summer and fall of 1984, I was in frequent contact with the panelists, assigning to each of them the responsibility for one or two facets of the material we would have to review prior to the Conference itself, which was scheduled for December 10–12 at the NIH Clinical Center in Bethesda, Maryland.

SPEAKERS AT THE SYMPOSIUM AND THEIR TOPICS

The Symposium was stimulating, the coverage was extensive, and without exception the speakers were outstanding authorities in their fields. Panel members, as usual at these Conferences, were ineligible for formal presentations.

The attendees enjoyed up-to-date surveys of the data bearing on the question at hand from outstanding leaders in the field. The program was as follows.

1. Evidence from pathology and animal models. Thomas B. Clarkson, D.V.M., Professor of Comparative Medicine, Bowman Gray School of Medicine.

2. Metabolic and genetic evidence: how LDL receptors influence cholesterol metabolism and atherosclerosis. Joseph L. Goldstein, M.D., Paul J. Thomas Professor of Genetics, University of Texas Health Science Center.

3. Evidence from prospective and other epidemiologic studies. Jeremiah Stamler, M.D., Professor and Chairman, Department of Community Health and Preventive Medicine, Northwestern University Medical School.

4. Does lowering blood cholesterol prevent heart disease? A critique of the evidence. Robert E. Olson, M.D., Ph.D., Professor of Medicine and Pharmacological Sciences, State University of New York, Stony Brook.

5. Summary of results from dietary and drug intervention studies. Richard Peto, M.Sc., Reader in Cancer Studies, Clinical Trial Service Unit, Radcliffe Infirmary, Oxford University.

6. Evidence from the Coronary Primary Prevention Trial. Basil M. Rifkind, M.D., F.R.C.P., Chief, Lipid Metabolism-Atherogenesis Branch, National Heart, Lung, and Blood Institute.

7. Relationship of clinical trial findings to epidemiologic data. Herman A. Tyroler, M.D., Professor of Epidemiology, University of North Carolina School of Public Health.

8. The nature of plasma cholesterol and the population distribution of cholesterol levels. Robert I. Levy, M.D., Professor of Medicine, Columbia University College of Physicians and Surgeons.

9. Efficacy of dietary management and associated risks. Scott M. Grundy, M.D., Ph.D., Professor of Internal Medicine and Biochemistry and Director, Center for Human Nutrition, University of Texas Health Science Center.

10. Maximal cholesterol lowering from diet. William E. Connor, M.D., Professor of Medicine, Oregon Health Sciences University.

11. Efficacy of drug management and associated risks. W. Virgil Brown, M.D., Professor of Medicine, Mount Sinai School of Medicine.

12. Identification and management of individuals with markedly elevated cholesterol. DeWitt S. Goodman, M.D., Tilden-Weger-Bieler Professor of Medicine, Columbia University College of Physicians and Surgeons.

13. Screening for hypercholesterolemia. Michael F. Oliver, M.D., F.C.R.P., Duke of Edinburgh Professor of Cardiology, University of Edinburgh.

14. What are the optimal cholesterol levels toward which we should aim for the American public at large? Antonio M. Gotto, Jr, M.D., D.Phil., Chairman, Department of Medicine, Baylor College of Medicine.

15. Identification and management of individuals with moderately elevated cholesterol. Edwin L. Bierman, M.D., Professor of Medicine and Head, Division of Endocrinology and Nutrition, University of Washington.

16. The appropriateness of public health measures to change American dietary habits to reduce blood cholesterol. Henry Blackburn, M.D., Professor and Director, Division of Epidemiology, University of Minnesota School of Public Health.

17. The lack of appropriateness (at this time) of public health measures to change American dietary habits. E.H. Ahrens, Jr., M.D., Professor, Rockefeller University.

18. The role of public education regarding cholesterol and heart disease. Kristen McNutt, Ph.D., J.D., Associate Director, Good Housekeeping Institute.

19. The role of the food industry. Walter M. Meyer, Associate Director, Food Product Development, Procter and Gamble Company.

20. What research directions should be pursued on the relationship between blood cholesterol and heart disease? Howard A. Eder, M.D., Professor of Medicine, Albert Einstein College of Medicine.

No attempt will be made to summarize the proceedings in detail except to say that the presentations were scholarly and there was lively discussion from the floor. Three points, however, deserve special comment.

First, Dr. Richard Peto from Oxford University (now Sir Richard), made an electrifying presentation that included what must have been one of his first applications of the method of meta-analysis, which he is generally credited with introducing into epidemiology. His analysis included all 17 of the appropriate diet or drug intervention studies available at the time. He concluded that lumping all the diet studies together there was a statistically significant decrease in coronary heart disease risk when blood cholesterol was lowered. He found the same to be true for the pooled drug intervention studies. Let me quote from his presentation: "It seems that about a 10 percent reduction in cholesterol is producing about a 20 per cent reduction in the incidence of cardiac events, that's first events, fatal or not." He himself never formally published this analysis but he had previously sent the data to two colleagues, J.I. Mann and J.W. Marr, and encouraged them to include it along with their review of intervention studies, which they published in 1981 (37). Meta-analysis is in essence a way of formalizing the dictum to "consider all of the evidence." Every scientist does this when he tries to decide if there is a consistent pattern in the data available from all the published papers on a given question. Peto's way of pooling results from several different studies in a formalized way with appropriate weighting offered an objective, quantifiable way of doing that. His meta-analysis of the intervention data had an important effect on the Panel's deliberations.

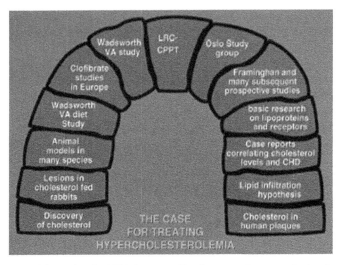

FIGURE 7.4   A slide used in the 1984 Consensus Panel presentation to emphasize the point that the Panel's conclusions were based not solely on the results of the CPPT but rather on all the many lines of available evidence, an "Arch of Evidence," to which the CPPT results added the all important keystone.

Second, there was ample opportunity for those who had strongly negative positions regarding the lipid hypothesis to voice their opinions and they took advantage of that opportunity. Three of the formal presentations were made by on-record skeptics with respect to intervention – Edward H. Ahrens, Jr, Michael F. Oliver, and Robert E. Olson.

Third, the panelists did not draw their conclusions on the basis of the CPPT results alone. As reviewed by the outstanding speakers, there was already an impressive body of evidence of varying kinds that supported the lipid hypothesis. The Panel emphasized this, using the slide shown in Figure 7.4 to characterize the CPPT results as the keystone in an arch of evidence supporting the lipid hypothesis.

THE CONSENSUS CONFERENCE AS PRESSURE COOKER

The Consensus Conference has been described, with some justification, as a "pressure cooker operation" and it has been implied that it is too hurried to allow for careful, informed judgment. What critics may not realize is that for months prior to the meeting in Bethesda the panelists are doing extensive research on topics assigned to them and exchanging ideas with each other by phone or by correspondence. The days in Bethesda, however, are hectic.

Our Panel members arrived in Bethesda on Sunday and, after a dinner at which Dr. Jacoby briefed us on the modus operandi, we got right down to work. We were still working at midnight. At 08:30 Monday the scientific program in the Clinical Center began and ran, with only a brief break for lunch, until 17:00. The Panel members had dinner together and again worked until close to midnight, comparing notes on the day's presentations and hearing reports from members assigned particular topics for in-depth research. On Tuesday morning at 08:30 we were back at the Clinical Center and listened to papers on the public health aspects of the problem until noon.

Then we went into Executive Session and *really* had to get down to work because we had to have a more-or-less final draft statement ready to present for review the following morning at 08:30. On the very first straw vote there was complete consensus on the first two questions: was blood cholesterol causal, and would reducing it help prevent heart disease? On these the Panel voted unanimously, yes.

Most of the next 11 hours were devoted to animated discussion of what constitutes a significantly high level of blood cholesterol and how it should be managed. Many panelists felt that the so-called "normal" cholesterol levels in the United States were much too high and that we should set a goal of 200 mg/dl (or even less) for everyone. However, we had to be realistic. A recommendation that seemed too radical might be a turn-off. Also we wanted to keep things simple. If we set up too many categories and there were too many numbers to keep in mind, the practicing physician, somewhat skeptical to begin with, would just tune us out. Robert W. Mahley came up with a simple set of numbers that satisfied everyone. We proposed "desirable" levels of less than 200 for persons under 20 years of age; less than 220 for those 30–39; and less than 240 for those above age 40; we proposed the same guidelines for men and women. The following year the National Cholesterol Education Program, discussed below, would issue more detailed guidelines but along similar lines (38). The other tough issue was what dietary recommendations to make. After much discussion we came up with guidelines very much like those adopted previously by the American Heart Association, namely, to exercise and reduce total calories so as to maintain normal body weight, to decrease total calories from dietary fat to 30 percent (less than 10 percent from saturated fat), and to reduce total daily cholesterol intake to less than 300 mg.

We also advised that, under the guidance of the National Heart, Lung and Blood Institute, there be established a national program, involving all of the major medical and public health associations, to educate both physicians and the public to the importance of controlling cholesterol levels. That program was officially launched the following year as the National Cholesterol Education Program (NCEP). Under the leadership of Dr. James I. Cleeman it has become a highly successful mechanism for providing education and guidance on management to physicians and patients alike (39;40).

RECEPTION BY THE PROFESSION AND THE PRESS

The report of the Consensus Conference was published in the *Journal of the American Medical Association* and in several other major journals at the beginning of 1985. By and large the reception was gratifyingly positive. Within a year or two the Europeans and the Canadians had proposed cholesterol guidelines that were quite similar, although differing slightly in cut-points and specific treatment recommendations. But the Cholesterol Wars were by no means over yet.

The most widely read general scientific journal, *Science*, covered the Conference but the published article was entitled "Heart Panel's Conclusions Questioned." It dwelt as much or more on the points of view of a handful of vocal dissenters as it did on the unanimous views of the expert panel and the supporting views expressed by the majority of the invited participants who spoke from the floor. Did the dissenters quoted in this *Science* piece have access to different data? No. Did they have broader and deeper relevant experience? No. Did they poke holes in the rationale by which the expert panel reached its conclusions? No. Did they actually represent a larger number of professionals in the field than did the expert panel? No. It is simply that dissent is always more newsworthy than consensus. This is especially true if the dissenters are highly vocal and even more so if they claim to be exposing flaws in the establishment position. If they can, in addition, imply malfeasance and conspiracy, so much the better. That's news.

To be sure science reporters are in a tough spot. In the absence of any formal polls, it is easy for convinced investigators to persuade them that their own views represent the majority position. The reporter from *Science*, Gina Kolata, was and is knowledgeable, accurate, and, justifiably, widely respected. But several highly credentialed experts were telling her that the conclusions of the Consensus Panel were wrong. She may have assumed that she was dealing with an issue that had two sides and tried to give them equal weight. What she probably did not know is that the small group that talked with her after the conference represented a rather small minority of the experts in the field. Most of the time reporters have no way of knowing which point of view represents the majority view. They can hardly be expected to take their own poll or contact all the relevant experts for a head count. In this particular case, however, just such a poll had already been conducted and the results published in 1978, though it was not paid much attention and the *Science* reporter was undoubtedly not aware of it. Dr. Kaare R. Norum, of the University of Oslo, conducted a survey to see whether there was a consensus among experts on atherosclerosis with respect to the role of blood cholesterol (41). He sent a questionnaire to 211 epidemiologists, nutritionists, geneticists, and others doing research on lipids and atherosclerosis. The list included almost every prominent researcher in the field at the time. Over 90 percent of those contacted responded, so the results were representative. In answer to the question, "Do you think there is a connection between plasma cholesterol level and the development of coronary heart disease?", 189 of those responding said "Yes", two said "No", and two were "Uncertain." To the question, "Do you think that our knowledge about diet and coronary heart disease is sufficient to recommend a moderate change in the diet for the population in an affluent society?", 176 of those responding said "Yes", 16 said "No", and one was "Uncertain." Norum's paper was lost sight of. It is true that his questionnaire did not probe deeply into the many complexities of the cholesterol/heart attack problem but it clearly showed that even 6 years before the NIH Consensus Conference over 90 percent of the experts in the field found the evidence linking blood cholesterol causally to heart attacks already very strong – strong enough to warrant recommendations that people should modify their diets to try to lower their blood cholesterol. Most recommended a decrease in saturated fat intake and a decrease in total calorie intake, exactly what the 1984 Consensus Conference recommended.

So here was an example of the fallacy of the oft quoted but misleading aphorism that "There are two sides to every story." The aphorism implies that the two sides both have about the same validity, that experts are more or less evenly divided on the issue, and that both sides are equally persuasive.

In this case there were indeed "two sides to the story" – cholesterol does or does not have anything to do with atherosclerosis – but the two sides were anything but equal. In Norum's survey, 189 of the experts said it does and only two of them said it doesn't! The reporter from *Science* was giving much too much emphasis to the views of the less than 1 per cent not yet convinced. *Science* accepted for publication a letter from me pointing out the skewed nature of the report and one from Dr. Jakoby on behalf of the Office of Applications of Medical Research, but follow-up letters are fairly ineffectual. Some damage had already been done.

A few months after the Consensus Conference, Michael F. Oliver published in the *Lancet* a piece entitled "Consensus or nonsensus conferences on coronary heart disease" (34). He wrote that: "The panel of jurists ... was selected to include experts who would, predictably, say ... that all levels of blood cholesterol in the United States are too high and should be lowered." Oliver was, and still is, a major figure in British cardiology who had been involved in many vanguard studies relating to coronary heart disease and atherosclerosis. His opinions needed to be taken into account. So how does the bedeviled reporter know how to weigh his strongly worded dissent? In my published response to Oliver's piece in the *Lancet*, which I entitled "Consensus Minus One?" I gently pointed out that the panel of 14 experts reached its conclusions unanimously and that "there were no more than a handful among some 600 conferees who appeared to disagree with the general terms of the recommendations." Oliver was himself an invited participant and duly had his say from the floor, as did two other well-known dissenters quoted in the *Science* article, Dr. Thomas Chalmers and Dr. Paul Meier.

Things got even worse. A few years after the Consensus Conference the *Atlantic* published and featured on its cover an article by Thomas J. Moore entitled "The cholesterol myth" (36). Moore, a journalist covering science, wrote: "the dissenters have been overwhelmed by the extravaganza put on not just by the heart institute but by a growing coalition that resembles a medical version of the military-industrial complex. This coalition includes ... the 'authorities' ... the heart institute [The National Heart, Lung and Blood Institute] itself ... and the American Heart Association." Moore then went on to name explicitly five investigators very active in the lipid field at the time who had offered to make themselves available to answer questions about the statins, which had just been introduced by Merck for clinical use (Antonio Gotto, Scott M. Grundy, John LaRosa, Robert I. Levy, and Daniel Steinberg). There followed a series of short paragraphs about this "Gang of Five" (my term) and the arguably actionable conclusion that: "It is likely that one reason these physicians consented to such an arrangement is that their laboratories were heavily involved in research funded by Merck." Finally, borrowing from Marc Antony, "There is no reason to doubt the honesty, sincerity, and expertise of any of these men." Yes, the cholesterol controversy has seen its share of vitriol.

THE POSITION OF THE FOOD AND DRUG ADMINISTRATION

The question of whether lowering cholesterol levels would reduce coronary event rates was one the FDA had already dealt with to some extent. For them this was a very pragmatic issue: should they approve drugs for marketing solely on the basis that they were safe and effectively lowered cholesterol levels? Or must the manufacturers first provide direct evidence that the cholesterol lowering actually reduced frequency of clinical events?

There were good reasons for the agency to be wary about approving drugs in this category. Two of the drugs in the Coronary Drug Project – D-thyroxine and estrogenic hormone (in men) – had not only failed to confer benefit but had actually increased mortality (17). Clofibrate, used in extensive clinical trials in Europe, had reduced nonfatal myocardial infarction but had proved to be toxic, increasing diseases of the gastrointestinal tract, including cancer, and marginally increasing overall mortality. Triparanol had been approved by the FDA for cholesterol lowering but proved to have serious side effects, including cataracts and hair loss, and was withdrawn from the market (42). Key personnel at Merrell were indicted by a grand jury but pleaded *nolo contendere*. A large number of lawsuits were filed, involving settlements of millions of dollars.

With this background of experience there was understandably little enthusiasm either in the pharmaceutical industry or at the FDA for new drugs to treat hypercholesterolemia. Nicotinic acid and cholestyramine were already on the market in 1984 but the package insert indications did not include cholesterol lowering. Nicotinic acid was for use as a nutritional supplement for its action as a B vitamin and cholestyramine was to be used in the management of biliary atresia. Of course many physicians were using them to lower blood cholesterol levels but they were prescribing off label. It was not until the late 1980s that the FDA decided that the evidence linking blood cholesterol levels causatively to coronary artery disease was strong enough to justify approval of cholesterol-lowering therapy without requiring the manufacturers to submit, at the time of application, clinical trial data demonstrating efficacy. According to Dr. Solomon Sobel, head of the FDA's Division of Metabolic and Endocrine Drugs at the time Merck's lovastatin came up for review, the results of the LRC-CPPT and the conclusions of the 1984 Consensus Conference figured very large in the FDA decision to begin approving drugs for the purpose of lowering blood cholesterol levels without requiring in advance a demonstration of efficacy against cardiac events. This change in policy smoothed the way for the later quick approval of the statins.

THE NATIONAL CHOLESTEROL EDUCATION PROGRAM

Armed now with the results of the Coronary Primary Prevention Trial and a consensus among the leaders in the field, the National Institutes of Health decided to go into high gear. It accepted for the first time the need to make lowering blood cholesterol levels a high priority goal. Implementation would be difficult and expensive but it would be well worth the effort and the cost. So, in 1985, the National Heart, Lung and Blood Institute Director, Claude Lenfant, took the lead by formulating plans for a national cooperative program to educate both health professionals and the public: the National Cholesterol Education Program (NCEP). A Coordinating Committee was established that included representatives from 24 important national health professional organizations, including the American Medical Association, the American Public Health Association, the American Heart Association, and so on. It also included representatives from ten other federal agencies. This was going to be full court press and everyone was going to be on the team. James I. Cleeman was put in charge of the program and has continued to run it effectively and imaginatively to this day (39;40).

One measure of the impact of the program is the increase in recognition by physicians of the importance of elevated cholesterol levels. The level of total cholesterol at which physicians would consider offering dietary advice fell from 260–279 mg/dl in 1983 to 200–219 in 1995. Between 1986 and 1995 the percentage of physicians who rated LDL as a very important marker for heart disease increased from 34 percent to 75 percent. Of course, the increasing press and television coverage over the years and doctor–patient interactions contributed as well, but the NCEP played a key role in this sea change in "cholesterol awareness."

Probably the most important contribution of the NCEP was to propose guidelines for diagnosis and treatment. At what level of cholesterol or LDL would dietary intervention be indicated? At what level would the use of drugs be warranted? This becomes a matter of judgment, balancing expected benefit against possible harmful effects and cost. The physicians in private practice, however astute, can hardly be expected to have all the facts at their fingertips. They need help. They need guidelines from experts who have studied the data on which such judgments rest. The NCEP took on the responsibility for providing those guidelines. At the very outset a panel was convened to flesh out the recommendations of the 1984 Consensus Conference, defining cut points at which diets or drugs should be used and setting treatment goals. In 1988, the first Adult Treatment Panel published its detailed guidelines (38), which quickly became the gold standard for who to treat and how to treat. Many other countries followed the U.S. lead and convened expert groups to develop their own guidelines. Except for relatively minor differences in cut points, these were remarkably similar to those of the NCEP.

There have been two revisions of the 1988 guidelines since, taking into account newer information about risk factors and advances in dietary and drug treatment. The NCEP will continue to monitor advances in understanding of heart attack risk factors and their treatment, modifying their guidelines periodically.

WHERE WE STOOD IN 1985

Returning to the post-Consensus Panel situation of 1985, were we home free? No. While there was finally acceptance that blood cholesterol was a significant causative factor in coronary heart disease and while a national program was in place to do something about it, the war was not over. There remained several critical questions. Treatment clearly would reduce heart attack rates and deaths from heart attacks, but would it reduce overall mortality? What, if any, were the hazards in lowering blood cholesterol? Was there any point in treating women? Was there any point in treating the elderly? How old is old? Would diabetic patients benefit from treatment? Would long-term treatment (decades) show up toxic effects not yet appreciated?

These questions were not yet adequately answered in 1985, but they would be rather quickly answered in the next two decades. That was thanks to the introduction of the statin drugs for lowering cholesterol levels, as discussed in Chapter 8.

NOTES

1 The Panel included Fredrickson, Chair; Edwin L. Bierman, Professor of Medicine, University of Washington; David H. Blankenhorn, University of Southern California; William Castelli, Framingham Heart Study, NHLBI; William E. Connor, University of Iowa; Gerald R. Cooper, Communicable Disease Center; Theodore Cooper, Director, NHLBI; Seymour Dayton, VA Medical Center, Los Angeles; Howard A. Eder, Albert Einstein College of Medicine; Ivan D. Frantz, University of Minnesota; William Fridewald, NHLBI; DeWitt S. Goodman, Columbia University; Frederick T. Hatch, University of California Berkeley; Richard J. Havel, University of California San Francisco; Peter Koo, University of Pennsylvania; Robert S. Lees, Massachusetts Institute of Technology; Robert I. Levy, NHLBI; Robert P. Noble, Sharon Research Institute; Isidore Rosenfeld, Cornell University; Daniel Steinberg, University of California San Diego.

2 The participating centers were at Baylor College of Medicine under Antonio M. Gotto; University of Cincinnati Medical Center under Charles J. Glueck; George Washington University Medical Center under John C. LaRosa; University of Iowa Hospitals under William E. Connor and, later, Francois Abboud and Helmut Schrott; Johns Hopkins Hospital under Peter O. Kwiterovich; University of Minnesota under Ivan D. Frantz, Jr. and Donald B. Hunninghake; Oklahoma Medical Research Foundation under Reagan H. Bradford; Washington University School of Medicine under Gustave Schonfeld; University of California San Diego under W. Virgil Brown and Daniel Steinberg and later Fred H. Mattson; University of Washington under William R. Hazzard and Edwin L. Bierman and later Robert H. Knopp; Stanford University under John W. Farquhar; Universities of Toronto and McMaster under J. Alick Little.

3 The original Intervention Committee appointed in 1971 included: Daniel Steinberg and John W. Farquhar, co-chairs; William R. Hazzard; Edmond A. Murphy; Al Oberman and Richard D. Remington; Dale Williams and James E. Grizzle as Data Coordinating Center representatives; and Robert I. Levy and Basil Rifkind, NHLI staff. Membership was later expanded to include William E. Connor; G. William Benedict; C.E. Davis; Ronald W. Fallat; Antonio M. Gotto; Richard C. Gross; Donald B. Hunninghake; John C. LaRosa; Maurice Mishkel; Gustav Schonfeld; L. Thomas Sheffield; Thomas F. Whayne, Jr.; and Richard J. Havlik representing the Program Office.

4 The Planning Committee was made up of Basil M. Rifkind, Chair; Susan Clark, Michael J. Bernstein, and Larry Blaser, representing the Office of Medical Applications of Research; Charles Glueck, Cincinnati General Hospital; William Hazzard, Johns Hopkins Medical School; Kenneth Lippel, Program Coordinator of the Atherogenesis Branch of NHLBI; Albert Oberman, University of Alabama Medical Center.

5 The Consensus Development Panel members were: Daniel Steinberg, M.D., Ph.D., Chair, University of California San Diego; Sidney Blumenthal, M.D., Columbia University; Richard A. Carleton, M.D., Brown University; Nancy H. Chosen, A.B., J.D., public interest attorney; James E. Dale, M.D., M.P.H., University of Massachusetts Medical School; John T. Fitzpatrick, Esq., Attorney at Law and past-President of the American Heart Association; Stephen B. Holley, M.D., M.P.H., University of California San Francisco; Robert W. Mahley, M.D., Ph.D., Director of

Gladstone Foundation Laboratories, University of California San Francisco; Gregory O'Keefe III, M.D., Islands Community Medical Center, Vinalhaven, ME; Richard D. Remington, Ph.D., University of Iowa; Elijah Saunders, M.D., University of Maryland School of Medicine; Robert E. Shank, M.D., Washington University School of Medicine; Arthur A. Spector, M.D., University of Iowa; and Robert W. Wissler, M.D., Ph.D., University of Chicago.

REFERENCES

1. 1984. The Lipid Research Clinics Coronary Primary Prevention Trial results. I. Reduction in incidence of coronary heart disease. *JAMA* 251:351–364.
2. 1984. The Lipid Research Clinics Coronary Primary Prevention Trial results. II. The relationship of reduction in incidence of coronary heart disease to cholesterol lowering. *JAMA* 251: 365–374.
3. Fredrickson, D.S., Levy, R.I., and Lees, R.S. 1967. Fat transport in lipoproteins – an integrated approach to mechanisms and disorders. *N Engl J Med* 276:34–42.
4. Fredrickson, D.S., Levy, R.I., and Lees, R.S. 1967. Fat transport in lipoproteins – an integrated approach to mechanisms and disorders. *N Engl J Med* 276:94–103.
5. Fredrickson, D.S., Levy, R.I., and Lees, R.S. 1967. Fat transport in lipoproteins – an integrated approach to mechanisms and disorders. *N Engl J Med* 276:215–225.
6. Gofman, J.W., Lindgren, F.T., and Elliott, H. 1949. Ultracentrifugal studies of lipoproteins of human serum. *J Biol Chem* 179:973–979.
7. Gofman, J.W., Lindgren, F., Elliott, H., Mantz, W., Hewitt, J., and Herring, V. 1950. The role of lipids and lipoproteins in atherosclerosis. *Science* 111:166–171.
8. Fredrickson, D.S. 1968. The field trial: some thoughts on the indispensable ordeal. *Bull N Y Acad Med* 44:985–993.
9. Diet–heart Review Panel of the National Heart Institute. 1969. Mass field trials of the Diet–heart question. American Heart Association, Inc.:1–37.
10. Buchwald, H., Varco, R.L., Matts, J.P., Long, J.M., Fitch, L.L., Campbell, G.S., Pearce, M.B., Yellin, A.E., Edmiston, W.A., Smink, R.D., Jr., Sawin, H.S., Jr., Campos, C.T., Nansen, B.J., Tuna, N., Karnegiss, J.N., Sammarco, M.E., Amplatz, K., Castaneda-Zuniga, W.R., Hunter, D.W., Bissett, J.K., Weber, F.J., Stevenson, J.W., Leon, A.S., and Chalmers, T.C. 1990. Effect of partial ileal bypass surgery on mortality and morbidity from coronary heart disease in patients with hypercholesterolemia. Report of the Program on the Surgical Control of the Hyperlipidemias (POSCH). *N Engl J Med* 323:946–955.
11. Buchwald, H., Varco, R.L., Boen, J.R., Williams, S.E., Hansen, B.J., Campos, C.T., Campbell, G.S., Pearce, M.B., Yellin, A.E., Edmiston, W.A., Smink, R.D., Jr., and Sawin, H.S., Jr. 1998. Effective lipid modification by partial ileal bypass reduced long-term coronary heart disease mortality and morbidity: five-year posttrial follow-up report from the POSCH. Program on the Surgical Control of the Hyperlipidemias. *Arch Intern Med* 158:1253–1261.
12. Dewar, H.A. and Oliver, M.F. 1971. Secondary prevention trials using clofibrate: a joint commentary on the Newcastle and Scottish trials. *Br Med J* 4:784–786.
13. Cooper, J., Geizerova, H., and Oliver, M.F. 1975. Letter: clofibrate and gallstones. *Lancet* 1:1083.
14. Oliver, M.F. 1978. Cholesterol, coronaries, clofibrate and death. *N Engl J Med* 299:1360–1362.
15. Hunninghake, D.B. 1991. Bile acid sequestrants. In *Drug Treatment of Hyperlipidemiai,* B.M. Rifkind, ed. Marcel Dekker, Inc., New York:89–102.
16. Pinckney, E.R. and Smith, R.L. 1987. Statistical analysis of lipid research clinics program. *Lancet* 1:503–504.

17. 1972. The Coronary Drug Project. Findings leading to further modifications of its protocol with respect to dextrothyroxine. The Coronary Drug Project Research Group. *JAMA* 220:996–1008.

18. 1971. Trial of clofibrate in the treatment of ischaemic heart disease. Five-year study by a group of physicians of the Newcastle upon Tyne region. *Br Med J* 4:767–775.

19. Scandinavian Simvistatin Survival Study Group. 1994. Randomised trial of cholesterol lowering in 4444 patients with coronary heart disease: the Scandinavian Simvistatin Survival Study (4S). *Lancet* 344:1383–1389.

20. Shepherd, J., Cobbe, S.M., Ford, I., Isles, C.G., Lorimer, A.R., Macfarlane, P.W., McKillop, J.H., and Packard, C.J. 1995. Prevention of coronary heart disease with pravastatin in men with hypercholesterolemia. West of Scotland Coronary Prevention Study Group. *N Engl J Med* 333:1301–1307.

21. Sacks, F.M., Pfeffer, M.A., Moye, L.A., Rouleau, J.L., Rutherford, J.D., Cole, T.G., Brown, L., Warnica, J.W., Arnold, J.M., Wun, C.C., Davis, B.R., and Braunwald, E. 1996. The effect of pravastatin on coronary events after myocardial infarction in patients with average cholesterol levels. Cholesterol and Recurrent Events Trial investigators. *N Engl J Med* 335:1001–1009.

22. 2002. MRC/BHF Heart Protection Study of cholesterol lowering with simvastatin in 20,536 high-risk individuals: a randomised placebo-controlled trial. *Lancet* 360:7–22.

23. Gould, A.L., Rossouw, J.E., Santanello, N.C., Heyse, J.F., and Furberg, C.D. 1998. Cholesterol reduction yields clinical benefit: impact of statin trials. *Circulation* 97:946–952.

24. Simons, L.A. 1984. The lipid hypothesis is proven. *Med J Aust* 140:316–317.

25. Podell, R.N. 1984. Coronary disease prevention. Proof of the anticholesterol pudding. *Postgrad Med* 75:193–196.

26. Oliver, M.F. 1984. Hypercholesterolaemia and coronary heart disease: an answer. *Br Med J (Clin Res Edn)* 288:423–424.

27. Mann, G.V. 1985. Coronary Heart Disease – "Doing the Wrong Thing." *Nutrition Today* 20 (Jul–Aug):12–14.

28. Collins, R., Keech, A., Peto, R., Sleight, P., Kjekshus, J., Wilhelmsen, L., MacMahon, S., Shaw, J., Simes, J., Braunwald, E., Buring, J., Hennekens, C., Pfeffer, M., Sacks, F., Probstfield, S., Yusuf, S., Downs, J.R., Gotto, A., Cobbe, S., Ford, I., and Shepherd, J. 1992. Cholesterol and total mortality: need for larger trials. *Br Med J* 304:1689.

29. Frick, M.H., Elo, O., Haapa, K., Heinonen, O.P., Heinsalmi, P., Helo, P., Huttunen, J.K., Kaitaniemi, P., Koskinen, P., Manninen, V., Maenpaa, H., Malkonen, M., Manttari, M., Norola, S., Pasternack, A., Pikkaraineen, J., Romo, M., Sjoblem, T., and Nikkila, E.A. 1987. Helsinki Heart Study: primary-prevention trial with gemfibrozil in middle-aged men with dyslipidemia. Safety of treatment, changes in risk factors, and incidence of coronary heart disease. *N Engl J Med* 317:1237–1245.

30. Virkkunen, M. 1979. Serum cholesterol in antisocial personality. *Neuropsychobiology* 5:27–30.

31. Virkkunen, M. and Penttinen, H. 1984. Serum cholesterol in aggressive conduct disorder: a preliminary study. *Biol Psychiatry* 19:435–439.

32. Wysowski, D.K. and Gross, P. 1990. Deaths due to accidents and violence in two recent trials of cholesterol-lowering drugs. *Arch Intern Med* 150:2169–2172.

33. Nestel, P.J. 1984. Time to treat cholesterol seriously. *Aust N.Z. J Med* 14:198–199.

34. Oliver, M.F. 1985. Consensus or nonsensus conferences on coronary heart disease. *Lancet* 1:1087–1089.

35. Kolata, G. 1985. Heart panel's conclusions questioned. *Science* 227:40–41.

36. Moore, T.J. 1989. The cholesterol myth. *Atlantic* (September).

37. Mann, J.I. and Marr, J.W. 1981. Coronary heart disease prevention. In *Lipoproteins, Atherosclerosis and Coronary Heart Disease*, N.E. Miller and B. Lewis, eds. Elsevier, Amsterdam:197–210.

38. 1988. Report of the National Cholesterol Education Program Expert Panel on Detection, Evaluation, and Treatment of High Blood Cholesterol in Adults. The Expert Panel. *Arch Intern Med* 148:36–69.

39. Cleeman, J.I. and Lenfant, C. 1998. The National Cholesterol Education Program: progress and prospects. *JAMA* 280:2099–2104.

40. Cleeman, J.I. 1989. National Cholesterol Education Program. Overview and educational activities. *J Reprod Med* 34:716–728.

41. Norum, K.R. 1978. Some present concepts concerning diet and prevention of coronary heart disease. *Nutr Metab* 22:1–7.

42. Fine, R.A. 1972. *The Great Drug Deception: The Shocking Story of MER/29*. Stein and Day, New York.

CHAPTER 8

Inhibition of Cholesterol Biosynthesis, the Discovery of the Statins, and a Revolution in Preventive Cardiology

LOWERING PLASMA CHOLESTEROL LEVELS BY INHIBITING ENDOGENOUS CHOLESTEROL BIOSYNTHESIS

Years before the causal relationship between blood cholesterol levels and coronary heart disease risk was widely accepted, there was already considerable interest in the possibility of using drugs to lower cholesterol levels, especially in patients with markedly elevated levels and strikingly high risk. A comprehensive 1962 review of activity in this field listed quite a few agents, some already in clinical use, but most still on the drawing boards (1). Practitioners had precious few choices available and the efficacy of what was available was limited. Because treatment once started would presumably be for a lifetime, the notion of using drugs at all seemed somewhat quixotic, but the notion of using drugs that would inhibit cholesterol biosynthesis seemed even more quixotic. Skeptics pointed out, and quite rightly, that the cholesterol molecule is crucially important as an essential component of all cell membranes and also as an obligatory precursor for the synthesis of steroid hormones and bile acids. Would not these vital functions be compromised, leading to unacceptable toxic side effects? Well, possibly so – especially if inhibition were to be complete or nearly so. But might it not be possible to titrate dosage so as to inhibit lipoprotein production without compromising those functions for which cholesterol was essential?

That was the gist of the proposal put forward in the early 1950s by Jean Cottet and his collaborators in France (2;3) and by Steinberg and Fredrickson

The Cholesterol Wars by D. Steinberg.
Copyright © 2007 Elsevier Inc. All rights of reproduction in any form reserved.

in the U.S. (4;5). However, neither group came up with an effective compound. The drug introduced by Cottet and co-workers (α-phenylbutyric acid) did slow the rate of incorporation of radioactive acetate into cholesterol but it did not actually decrease *de novo* production of cholesterol molecules. That is because the compound inhibited the activation of free acetate to acetyl coenzyme A (6) but none of the later steps in cholesterol synthesis. Actually the activation of acetate is not essential for endogenous cholesterol biosynthesis, a point often lost sight of. The degradation of all major foodstuffs (fatty acids, carbohydrates, and some amino acids) generates acetyl coenzyme A, not free acetate. The acetyl coenzyme A deriving from any of these pathways can then serve directly as the starting point for cholesterol synthesis, i.e. without being first degraded to free acetate. Therefore inhibition of the conversion of acetate to acetyl coenzyme A would not necessarily compromise net cholesterol production – and indeed it did not. The reported cholesterol-lowering effects of Cottet's compound in animals and in humans could not be confirmed (7–9).

Steinberg and Fredrickson, following up on observations made by Tomkins, Sheppard, and Chaikoff at Berkeley (10), confirmed that a close chemical relative of cholesterol, delta-4-cholestenone, could inhibit cholesterol synthesis and went on to show that it reduced blood cholesterol levels (5;11). However, feeding the compound caused accumulation of cholestanol (12), which was itself proatherogenic, and toxic side effects of the compound precluded clinical use (11). These early efforts failed to solve the problem but they did at least spark interest in the possibility that cholesterol synthesis might be a legitimate pharmacologic target.

THE MER/29 (TRIPARANOL) SCANDAL: A SETBACK IN THE QUEST FOR DRUGS INHIBITING CHOLESTEROL BIOSYNTHESIS

In the mid-1950s, while randomly screening their chemical library for compounds that might lower blood cholesterol levels, the Wm S. Merrell Company came across a compound that looked very promising. In rats and in dogs it appeared to lower cholesterol levels by as much as 20–25 percent. It was assigned the in-house identifier "MER/29" and the generic name of triparanol, under which it was later marketed. The company demonstrated that the drug was an inhibitor of cholesterol biosynthesis (13). Later studies at the NIH established the exact site of action of the drug – it inhibited the very last step on the synthetic pathway, i.e. the conversion of desmosterol to cholesterol (14), by reduction of the side-chain double bond. Knowing the site of inhibition led to a number of findings that challenged the value of triparanol. For example, it was shown that large amounts of desmosterol accumulated in the plasma of treated animals and patients, accounting for up to 30 percent of total plasma sterols (15). A key point of confusion was that the color yield of desmosterol in the then standard Liebermann-Burchard reaction was only about half that of cholesterol. Consequently the total sterol level in the blood (the sum of cholesterol and desmosterol) was being underestimated and the apparent drop in cholesterol level overestimated. Furthermore, it was shown that desmosterol entered athero-sclerotic lesions just as effectively as cholesterol, not surprisingly since it is structurally different from cholesterol by just one double bond (16). So even though triparanol did have a modest effect in lowering blood cholesterol levels, that effect was significantly less than it appeared to be; moreover, the accumulating precursor would probably substitute nicely for cholesterol in atherogenesis.

Worst of all, in addition to being relatively ineffective, the drug had serious toxic side effects. It was quickly found that it caused lens cataracts and hair loss in rats and dogs. Rats on high doses actually became blind. Investigators working on the mechanism of action of the drug were well aware of these toxic effects and called them to the attention of the drug company. A group at Merck, Sharp and Dohme formally notified Merrell of these toxic effects several months before the drug was approved by the Food and Drug Administration (FDA) and invited a team from Merrell to visit their laboratories and see for themselves. The invitation was accepted and three Merrell people visited in January 1961. During the discussions the Merrell representatives denied having ever encountered cataracts in their own studies but indicated that they would "try to confirm the Merck experiments." It later emerged that Merrell toxicologists had in fact already observed eye damage and even blindness in some of their rats and dogs but failed to include that information in the material they submitted to the FDA. This omission would turn out to be a major factor in the 1963 federal grand jury criminal indictment brought against the company and some of its employees. The company pleaded *nolo contendere*, which protected them against the use of the grand jury findings in subsequent civil suits. Several hundred such suits were filed and these were settled by Merrell at a cost of about $50,000,000. Hard data are not available, but it would not be surprising if the company netted that much during the year they kept the drug on the market – did so even in the face of the ever-increasing evidence that patients were developing serious eye problems and that the drug was not really lowering the concentration of blood sterols very much.[1]

IMPACT OF THE TRIPARANOL DEBACLE ON THE WAR AGAINST CHOLESTEROL

The triparanol fiasco caused many companies to bring to an abrupt halt their hunt for drugs that might block cholesterol synthesis even though the mechanisms underlying the lens damage and hair loss were not known. Conceivably the toxic effects might have been due to the cholesterol lowering per se and, if so, other inhibitors would then be expected to cause similar side effects. Alternatively, the toxic effects might be unique to triparanol (which turned out to be the case), so other inhibitors might be free of such effects. However, even if a pharmaceutical house should hit on an inhibitor that did not cause cataracts, there would be the problem of getting FDA clearance. The FDA would predictably now be much tougher and might require, as well they should, much more rigorous safety and efficacy data before approving any drugs in the class. The $50,000,000 it cost Merrell to settle the civil lawsuits against it surely helped other companies make the decision to "abandon ship." Fortunately there were some drug companies that did not completely shut down their research programs on inhibitors of cholesterol biosynthesis. However, most of them decided to steer clear of inhibitors that blocked synthesis at the later stages, hoping that inhibitors working at earlier steps would not share with triparanol the disastrous effects on the eye and on hair growth. Sad to say, for almost 20 years they came up with nothing usable. Between 1959, when triparanol was introduced, and 1979, when the statin drugs reached the clinic, there were many patent applications for inhibitors of cholesterol synthesis. Not one of them reached the clinic.

THE BIRTH OF THE STATINS: AKIRA ENDO[2]

Pharmaceutical companies in the 1970s were panning for antibiotic gold, systematically screening compounds made by fungi and other microbes for their potential as antimicrobial agents. They were inspired of course by Fleming's discovery of penicillin when one of his Petri dishes sat around too long and got contaminated by a fungus (21). The bacteria originally seeded onto his dish had grown thickly everywhere except for neat, clear circles surrounding the intrusive fungal colonies. Having a prepared mind, Fleming realized that the fungus was making something that killed bacteria in its immediate vicinity, a property that might be put to good use! His discovery was serendipitous, but soon the search for microbial antibiotics became systematized and was being pursued on a large scale.

In 1971, Akira Endo, working at the Sankyo Co. in Tokyo, speculated that the broths in which fungi were being grown in the hunt for new and better penicillins just might contain also natural inhibitors of cholesterol synthesis. There was at the time no direct evidence to support that speculation, but Endo states that he hoped that some microorganisms might "produce such compounds as a weapon in the fight against other microbes that required sterols or other isoprenoids for growth" (19). Parenthetically, it should be noted that

Endo's interest in cholesterol metabolism dated back at least to 1965, when he applied for a fellowship to work at Harvard with Konrad Bloch. Unfortunately Bloch had no fellowship openings available at the time and Endo went instead to New York where he spent 2 years as a Fellow at the Albert Einstein School of Medicine working in the laboratory of Bernard L. Horecker. On returning to Tokyo, Endo and his associate at Sankyo, Dr. Masao Kuroda, began to test fungal broths for their ability to inhibit cholesterol synthesis from labeled acetate in a cell-free system. The assay was straightforward, fast and cheap. Endo and his colleagues began testing in 1971. Week after week, month after month, they patiently applied their assay to these broths but the results were uniformly and depressingly negative. Two years and over 6,000 tests later, they finally came up with a real winner. The culture broth from *Penicillium citrinium* contained a remarkably potent inhibitor of cholesterol synthesis (22;23), which they isolated and designated ML-236B. They showed that it inhibited incorporation of acetate but not that of mevalonate into cholesterol. They pointed out that the ML-236B molecule included a domain homologous to hydroxymethylglutarate and thus the presumptive site binding it to the reductase. ML-236B, for historical reasons discussed below, was referred to in the early years as *compactin* and the name stuck. We shall continue to refer to it that way in this review.

So now Sankyo had a specific inhibitor working at the HMGCoA reductase step. The next question was whether it would work *in vivo* and whether it would be tolerated at effective dosages.

Endo's first tests were done in rats using just single doses, probably because the amounts of compound available were limited. It seemed at first to work but when given in repeated doses over a longer period of time there was no consistent effect on blood cholesterol levels (24). It looked as if 2 years of work and over 6,000 tests had led nowhere. Fortunately, Endo and associates did not give up at this point, as they might well have done. They went on to try their compound in dogs, and there the results were quite different; now they saw a very significant and consistent lowering of blood cholesterol levels (25). They also showed that the drug worked in rabbits, hens, and monkeys (26). In retrospect, the reason for the initial "failure" in rats is clear. The drug does actually inhibit cholesterol synthesis *in vivo* in the rat, just as effectively as it does in other species, even though there is a compensatory increase in the amount of reductase enzyme. However, rats have extremely low LDL levels. Most of their plasma cholesterol is in the HDL fraction. Consequently even a significant percentage reduction of LDL might not show up as much of a reduction in total cholesterol level, which is what was measured in these early studies.

Endo's results did not draw a lot of attention initially. Partly this apathetic reception may have reflected the reaction to the triparanol fiasco reviewed above. There was no great enthusiasm in the pharmaceutical industry for another inhibitor of cholesterol biosynthesis in the 1970s. In 1977, Endo presented a paper in Philadelphia at a symposium on Drugs Affecting Lipid Metabolism, a triennial meeting to which all the major pharmaceutical companies sent representatives. Surprisingly his presentation was poorly attended. However, the exciting possibilities of compactin were not lost on Michael S. Brown and Joseph L. Goldstein at the University of Texas Southwestern Medical School (20). Within a month of the publication of Endo's first report on compactin they had written to Endo to ask for a sample to use in their ongoing studies of the regulation of cholesterol biosynthesis. Endo sent the samples and they invited him to visit them in Dallas after the Philadelphia meeting. They compared notes on their experiments done independently in Tokyo and in Dallas, found them to be concordant, and published the results jointly in a 1978 paper in the *Journal of Biological Chemistry* (27). This was an important paper because it described for the first time the huge increase in the amounts of the reductase enzyme induced in cells by statin treatment. Since the statins are competitive inhibitors, the inhibition, which is powerful at the drug concentrations reached within the intact cells, is largely lost when the tissue is homogenized and the cytoplasm greatly diluted for measurement of enzyme activity. These studies were done using human fibroblasts but the same phenomenon was later reported in hepatocytes. A few years later the Goldstein/ Brown laboratory showed that this huge overproduction of reductase protein, representing an attempt by the cell to overcome the statin inhibition, is accompanied by a huge build-up of endoplasmic reticulum, the organelle in which the reductase resides (28). As a result the cells look "abnormal" but of course they are not cancer cells. As discussed below, this may in retrospect be what led the pathologists at Sankyo at a later date to conclude, incorrectly, that high doses of compactin were possibly carcinogenic.

THE EARLY CLINICAL TRIALS OF COMPACTIN

In 1980, Yamamoto, Sudo, and Endo reported that compactin, given by mouth at a dose of 50 mg per day, decreased cholesterol levels in patients with hypercholesterolemia by an average of 27 percent (29). In some patients the drop was as much as 30–35 percent. A second clinical study in seven patients with heterozygous familial hypercholesterolemia, which is much harder to treat, was later published in the prestigious *New England Journal of Medicine* by Mabuchi *et al.* (30). It showed a highly significant drop in total cholesterol levels from 390 down to 303. There was no doubt now that, barring the possibility of some unsuspected toxicity showing up in larger and longer clinical trials, this drug and others like it were going to be wonder drugs. Akira Endo had inaugurated the statin era (Figure 8.1).

FIGURE 8.1   Akira Endo, the discoverer of the first statin drug in 1976 (22). Photo courtesy of Dr. Endo.

AN INSTRUCTIVE FOOTNOTE TO THE DISCOVERY OF THE STATINS

By a most remarkable coincidence, a British group at Beecham Laboratories in the United Kingdom, while searching for antibiotics, independently isolated precisely the same compound as Endo's ML-236B from a different but related mold and at almost the same time (31). They named it compactin. However, the Beecham workers were narrowly focused on antibiotics, and in that arena compactin was not particularly effective and so it was dropped from the Beecham program. They did not appreciate until later, after Endo had published on its potency as an HMGCoA reductase inhibitor, what a treasure they had had in their hands. Their initial report did not mention the close homology between the lactone ring in compactin and that of 3-hydroxymethylglutarate. Only in 1980, after Endo had already published several papers on the potency of compactin as an inhibitor, did the Beecham group explore compactin's effects on cholesterol biosynthesis *in vitro* and in rats *in vivo*. They readily confirmed Endo's results – it was a potent inhibitor (32). However, like Endo, they observed no change in blood cholesterol levels in rats despite a very respectable 70 percent inhibition of the rate of *in vivo* cholesterol synthesis. They concluded, quite incorrectly, that if the drug didn't work in rats, it probably would not work in other species, even though they noted in their discussion that Endo's group had already reported significant lowering of blood cholesterol levels in dogs and monkeys. What they seemed to be unaware of was that most of the blood cholesterol in rats is in the HDL fraction, so that even a significant lowering of LDL might go undetected. They suggested that the cholesterol lowering Endo had observed in dogs and monkeys might be due not to inhibition of cholesterol synthesis but some independent, unrelated effect of the drug. They concluded that trying to decrease blood cholesterol by inhibiting cholesterol synthesis was "futile" and dropped the project. Had they gone on and tested it in other species, as Endo had, Beecham might have entered the statin race early on.

MERCK ENTERS THE RACE: ALFRED W. ALBERTS AND P. ROY VAGELOS[3]

The dramatic clinical findings with compactin, even though limited to a small number of subjects, made quite a stir. Every pharmaceutical company of size soon began screening their microbial cultures not just for antibiotics but also for inhibitors of cholesterol biosynthesis. Merck, Sharp and Dohme was first out of the gate. Shortly after Endo's papers appeared, P. Roy Vagelos, President of Merck Research Laboratories, signed a confidentiality agreement with Sankyo and obtained samples of compactin. They quickly confirmed Endo's findings and were astonished at the potency of the drug. Under the direction of Alfred W. Alberts, a long-time Vagelos collaborator who had come with him from Washington University, Merck set out to find their own statin (Figure 8.2).

Alberts' group started screening in October 1978 and was lucky enough to hit pay dirt with sample No. 18, just 2 weeks into their program (34). Quite a contrast to Endo's experience: he screened about 6,000 broths before making a hit!

Alberts' lovastatin had a structure only differing by one methyl group from that of compactin and it had very similar biological properties. Preliminary clinical studies were begun in 1980 and the early results looked very promising indeed. But the whole Merck program was suddenly shut down in the fall of 1980. The story behind that is an intriguing one but we need to preface it by going back to Japan and the early work of Endo.

In 1979, Endo was offered an Associate Professorship at Tokyo Noko University and left Sankyo. He continued his pursuit of reductase inhibitors and in August of the same year he reported the isolation from cultures of a different fungus (*Monascus ruber*) another highly effective inhibitor of cholesterol synthesis, which he named monacolin K. Its chemical structure was very similar to that of compactin, differing only by the addition of a single carbon atom on one of the rings. He applied for a patent in Japan in February 1979.

Meanwhile, Merck was plowing ahead with its own screening program and, as mentioned above, very quickly hit their first promising inhibitor, a compound secreted by a fungus (*Aspergillus terreus*) quite distinct from the one Endo had used. Merck named their compound mevinolin (later changed to lovastatin) and applied for the U.S. patent in June, 1979. The remarkable fact is

FIGURE 8.2   Alfred W. Alberts, who headed the Merck group that isolated mevinolin (lovas-tatin), the first statin to reach the market and the one to be used in the first critical large-scale clin-ical trials (34). Photo courtesy of Dr. Alberts.

that the structures of Endo's monacolin K and that of Alberts' lovastatin turned out to be absolutely identical ... precisely the same compound produced by two different microbes and discovered independently in two different laboratories almost simultaneously! Endo and his university originally held the patent on monacolin K/lovastatin in Japan but later sold it to Sankyo. Merck held the patent on lovastatin in the United States but did not have worldwide rights. Sankyo was now already quite far along with their clinical studies on com-pactin, had published several papers on its use in man, and were probably going to market it any day. Merck was putting every effort into its lovastatin program and had already carried out a few clinical studies. The groundwork was now laid for a knockdown, drag-out race to see who would be the first successfully to bring a statin to market ... but then something strange happened.

HOW WE ALMOST LOST THE STATINS

In the fall of 1980, Merck held its annual 4-day research "retreat" at which each working group presented its most recent results and its plans for the coming year. That year it was held at the Seaview Resort at Absecon, New Jersey. P. Roy Vagelos, President for Research, had been driven down that morning for the meeting and he recalls very vividly the dramatic events of that afternoon. Merck was in excellent financial condition (net income about $400,000,000) thanks to an innovative drug discovery plan that Vagelos had initiated. That plan had brought several "blockbuster" drugs to the market over the preceding few years. Still, Vagelos knew that maintaining Merck's leadership role required that there be a continuing input of new products into the "pipeline." As Vagelos puts it, "[that's] why we were all watching *Mevacor* [lovastatin] so closely and that's why we were all so upbeat about our research program. We thought *Mevacor* had the potential to become a billion-dollar-a-year product" (33). The day's discussions went well and spirits were high.

Then, toward the end of the day, right in the middle of a wrap-up session, Vagelos was called out to take an urgent phone call from Japan. The call was from the head of Merck's Japanese research office, H. Boyd Woodruff. Sankyo had just terminated all its clinical studies with compactin. They had given no reasons for this startling move and were unwilling to answer any questions. Woodruff said, however, that rumors were circulating to the effect that the company had discovered intestinal lymphomas in their dogs treated with very large doses of compactin over a long period of time. Woodruff had tried to verify the rumors but the company would not comment. No one seemed to know what was going on. But one thing seemed certain: Sankyo would never have aborted a multimillion dollar program unless they had encountered something really ominous.

What had been a warm and comfortable, even self-congratulatory, company retreat suddenly became something of a wake. Lovastatin only differed in structure from compactin by one carbon atom. If compactin was carcinogenic, it was likely that lovastatin would be also. On the other hand, the carcinogenicity that had allegedly been encountered might be related not to the cholesterol-lowering effect per se but to an unrelated effect of the compactin molecule. Conceivably the one extra carbon on lovastatin might abolish any carcinogenic potential. However, that was a long shot. Merck was already carrying out studies on the effects of lovastatin in dogs and had not encountered any intestinal cancers, but their studies were so far of fairly short duration. Longer exposures might confirm the Japanese findings. Merck had already invested millions of dollars on this project. Halting it would mean losing months or years in the race to get their statin drug on the market. Alberts, who had discovered lovastatin, was devoting his energies full time to this project. Jonathan Tobert was well along with safety and efficacy testing in the clinic. Vagelos knew that this might

FIGURE 8.3   P. Roy Vagelos, President for Research at Merck at the time mevinolin was discovered and later CEO of the entire corporation. When rumors surfaced that Sankyo's compound might be carcinogenic he acted responsibly, i.e. he temporarily halted all clinical trials. Careful, long-term toxicity studies at Merck never uncovered any carcinogenic potential and mevinolin got the green light from the FDA in 1987.

be a real blockbuster drug, and their teams would be devastated if their project was junked. What to do?

Vagelos (Figure 8.3) did the right thing. He would not take any chance of exposing even one patient to a potential carcinogen, no matter what it might cost Merck and even if it meant they lost the race to be the first to bring a statin to market. He immediately called a halt to all clinical studies and asked investigators to return outstanding samples; he notified the FDA, and he decided to make an all-out effort to get to the bottom of the cancer rumors. At this point only a small number of patients had received lovastatin and only at low dosages, but still Merck advised their physicians to check carefully for any signs of cancer. None was found, either at that time or later, even after many years of testing in many thousands of patients all over the world. But in the fall of 1980 at Absecon, NJ, none of this was known yet and the mood was somber.

Merck had only rumors to go on and those rumors were unconfirmed and lacking in detail. How common were these tumors in dogs? At what dosage did they occur? How did that dosage compare to the dosage needed to treat human hypercholesterolemia? Alberts and some of the other Merck investigators wanted to continue at least the animal toxicity studies. A second group favored dropping the whole project and instead making every effort to find a different statin that would hopefully be totally "clean" with respect to carcinogenicity.

Vagelos tried every way he could to get more information about the findings that prompted Sankyo to drop its clinical trials, including letters and phone calls to the company's executives asking them to share the results of their safety assessment tests. Sankyo was unwilling to comment. Vagelos did get second-hand confirmation through an American pharmaceutical company that was working with the Japanese that the rumors might be true. So he and Barry Cohen, who was in charge of Merck's international businesses, went to Japan and visited Sankyo themselves. Vagelos offered a business deal: "If you help us solve this problem, we'll share *Mevacor* [lovastatin] with you in Japan and you can share your second-generation product with us when you're ready." The head of Sankyo smiled and said he would like to cooperate but that there were "others" who objected. Vagelos returned empty handed, puzzled – and angry.

THE IMPLICATIONS IF LOVASTATIN HAD BEEN DROPPED

Vagelos was now getting input from investigators who dealt directly with patients with heterozygous familial hypercholesterolemia, patients who could have fatal heart attacks as early as age 30. None of the drugs then available was very effective in these patients. The clinical studies in Japan, while limited in scope, had already shown that the cholesterol levels in these patients fell on compactin treatment and so for them the drug could be life saving. Even if there were risks, even life-threatening risks, those had to be balanced against the potential life-saving benefits of treatment. Dropping the program might be denying some patients a chance to prolong their lives. Moreover, there was still no hard evidence that compactin would be toxic in humans; there were only rumors about toxicity in dogs given very high doses. There was no way to be certain about the extent of risk, but the potential benefit could be great in view of the known natural history of coronary heart disease in such families.

This line of argument was urged on Vagelos by a number of clinicians who had used lovastatin in early safety and efficacy trials, including Roger Illingworth of Portland, and Scott M. Grundy, David Bilheimer, Joseph L. Goldstein, and Michael S. Brown of Dallas. Two members of Vagelos's Scientific Advisory Board, Jean Wilson from Dallas and Daniel Steinberg from La Jolla, made the same case. In his memoir Vagelos remembers, "we needed advice from the type of authorities in their field whom the FDA would consult."

Following these meetings Merck presented all of their data to the FDA and got a quick green light for additional clinical trials in high-risk patients. The request for clearance was submitted in January 1987 and approved in August the same year. The only qualification was that all patients be carefully evaluated for cataracts before starting treatment and be carefully followed because dogs on very high dosages (more than 100-fold the proposed human dosage) had developed cataracts. Large-scale studies showed no effect on cataract development and the requirement was withdrawn. Notably the FDA approval to go to market was not qualified or dependent on a commitment to demonstrate explicitly effectiveness against coronary heart disease events. The FDA position was that the NIH Coronary Primary Prevention Trial had closed the link between cholesterol lowering and reduction of coronary heart disease risk. The FDA approved lovastatin as a cholesterol-lowering drug and Merck was on its way to putting the first statin drug into the hands of clinicians.[4]

WAS COMPACTIN INDEED CARCINOGENIC?

The chronic toxicity studies of compactin at Sankyo were done using astronomically high doses – up to 200 mg/kg/day. Yamamoto, Sudo, and Endo (29) had already shown that the dosage needed to lower cholesterol levels even in severely hypercholesterolemic patients was less than 1 mg/kg/day. In other words, the dogs were getting about 200 times the dosage that would be used in patients. Still, the toxicologists at Sankyo felt obliged to counsel against continuing the use of even small doses being used in patients. Like most Japanese pharmaceutical houses, Sankyo was strongly tilted toward conservatism in the 1960s, partly because of several serious instances of post-marketing toxicity, including the tragic experience with thalidomide. This conservatism tended to be shared generally by the medical profession in Japan at the time. One prominent Japanese clinician warned, "powerful drugs, like a sharp knife, should be considered dangerous." Another warned students not to prescribe drugs at full dosage, thereby running the risk of toxicity, but whenever possible to use half the normal dose. Another factor may have been the somewhat parochial approach of the pharmaceutical companies in Japan at that time, an unwillingness openly to exchange information with and seek advice from those outside the company "family."[5] In any case Sankyo dropped compactin and continued to hunt for other fungal inhibitors.

In retrospect, we can now say with absolute confidence that neither lovastatin nor any other of the statin drugs on the market is carcinogenic, either in experimental animals or in humans. Clinical trials in which tens of thousands of subjects have received either a placebo or a statin have shown no change at all in cancer incidence. In the early 1980s, however, the level of anxiety both at Sankyo and at Merck was high – and we came close to losing these wonder drugs.

THE MIRACULOUS POWER OF THE STATINS
TO PREVENT HEART ATTACKS AND SAVE LIVES

It is difficult to overstate the impact the statins had on the management of atherosclerosis, particularly coronary heart disease and stroke. First, because the statins lowered blood cholesterol so much more than any of the existing diet or drug treatments, it suddenly became much easier to demonstrate the decrease in coronary heart disease events and to do so in a statistically significant, unarguable way. For example, in the ground-breaking 1984 NIH Coronary Primary Prevention Trial, using the drug cholestyramine (35;36), blood cholesterol in the treated group fell by only about 10 percent and LDL cholesterol by about 20 percent. This was enough to reduce the heart attack rate but only by about 20 percent. The result barely reached statistical significance. By contrast, in one of the first large-scale statin trials total cholesterol was reduced by 25 percent, LDL cholesterol by 35 percent, and coronary heart disease deaths by 42 percent! This reduction was significant at the $p < 0.00001$ level! This trial, the so-called 4S study in Scandinavia (37), was done using simvistatin, the second Merck statin, which was discovered while they were in the process of assessing the safety of lovastatin. The 4S study showed, for the first time in any cholesterol-lowering trial, a significant decrease in *all-cause* mortality. A new era in treatment of hypercholesterolemia and coronary heart disease had arrived.

A recent meta-analysis of 14 statin trials with an astonishing total of 90,056 individuals randomized (using lovastatin, simvistatin, pravastatin, fluvastatin, or atorvastatin) showed that the decrease in coronary events was best predicted by the *absolute* decrease in LDL levels. Incidence of major vascular events was reduced by about 20 percent for each 1 mM/L (40 mg/dl) drop in LDL cholesterol (38). Thus, an individual starting with an LDL of 280 mg/dl and who dropped that level down to 200 mg/dl on therapy (a 29 percent drop), would be predicted to have a 40 percent decrease in risk over a 5-year period.

Is the effect of the statins on coronary event rates all due to the extent to which they lower LDL? Or do their so-called pleiotropic effects (i.e. effects independent of cholesterol lowering) also play a significant role? There is no doubt that statins do have additional biologic effects beyond the lowering of LDL levels. These include anti-inflammatory effects, stabilization of endothelium, and decreasing oxidative stress (39). Some of these effects may be traced back to inhibition of protein prenylation or inhibition of ubiquinone and dolichol synthesis (i.e. effects of the primary inhibition at the HMGCoA reductase step) but independent effects at other sites of action are not ruled out. What is still not clear is whether and to what extent such pleiotropic effects contribute to the decrease in morbidity and mortality seen with statin treatment (40). The analysis by Gould *et al.* (41) offers an estimate based on calculating the extent of reduction in coronary heart disease mortality in all the trials preceding the introduction of the statins (both diet and drug intervention studies) and then adding the results with the statins for comparison. From the slope of the curve relating coronary heart disease mortality to the percentage lowering of cholesterol level in the pre-statin era, they estimated a 13 percent reduction for each 10 percent drop in cholesterol; after adding in the statin data, that number was 15 percent. Total mortality in the pre-statin era was reduced 10 percent for each 10 percent drop in cholesterol level; adding in the statin data, the number was 11 percent.

Second, the large-scale statin studies laid to rest the lingering concerns that lowering blood cholesterol levels might be intrinsically dangerous. This concern arose originally because in the European clofibrate trials there were indeed more deaths in the drug-treated group than in the controls, although the difference was marginal (42). In retrospect this difference was probably attributable not to the lowering of the cholesterol level per se but to a toxic effect intrinsic to the clofibrate molecule and unrelated to its cholesterol-lowering activity. The second generation fibric acids (e.g. gemfibrozil and fenofibrate) have not shared the toxicity of clofibrate (43;44). Concerns that lowering blood cholesterol levels might be intrinsically dangerous were misplaced a priori:

1. because levels of *intracellular* cholesterol are jealously guarded by the LDL receptor homeostatic mechanism (45); and
2. because most animal species have LDL levels well below those reached during even the most aggressive treatment of hypercholesterolemia (46).

Obviously these animals' cells do just fine. Nevertheless, this had been a concern and had deterred some physicians from treating hypercholesterolemia vigorously. The large-scale statin trials showed that even lowering LDL values to below 100 mg/dl was not only safe but actually decreased overall mortality significantly (37;38;47–51).

Third, there had been concern that while treating hypercholesterolemia might reduce coronary heart disease risk, it might at the same time lead somehow to increases in mortality from other causes, not necessarily because of the lowering of cholesterol levels per se but possibly due to metabolic dysfunctions arising from other properties of the cholesterol-lowering agents. Indeed, in the Coronary Primary Prevention Trial there had been a statistically significant decrease in coronary heart disease mortality and yet no decrease in all-cause mortality. Of course, as was pointed out at the time, the study was not designed to have the power to show a decrease in all-cause mortality; that would have required a larger number of subjects (35;36). Nevertheless, much was made of a small, statistically *non-significant* increase in the category of "violent deaths," which included suicides, homicides, and traumatic deaths (e.g. automobile accidents). (How homicides could be made more likely by the *victim's* cholestyramine intake was never made clear.) In any case, as first shown in the 4S trial using simvastatin (37) and borne out in the meta-analysis of over 90,000 subjects in 14 statin trials (38), statins *decreased* all-cause mortality.

One cause for concern about the safety of cholesterol lowering came from prospective epidemiologic studies. These showed that individuals with low blood cholesterol levels when initially surveyed, e.g. below 160 mg/dl, were more likely to die over the next 5 years than those with average cholesterol levels. In retrospect this was probably due to the fact that a number of potentially life-threatening diseases are characterized by low blood cholesterol levels in the early, pre-clinical stages of the disease. This is true, for example, in many forms of cancer and in cirrhosis of the liver. In other words, the poor prognosis in the group with initially low cholesterol levels might be accounted for by the fact that some fraction of them entered the study already ill. The NIH, in 1990, convened a panel of experts to discuss the possibility that lowering cholesterol levels might be intrinsically dangerous. They concluded that the evidence did not justify such a finding but with the data available at that time neither could they rule it out absolutely (52). The large-scale statin studies settled the issue. It is now clear that the marginal effects on all-cause mortality seen in the early trials was attributable in part to the small sizes of the populations studied and in part to the modest lowering of cholesterol levels.

Fourth, the large-scale statin studies made it clear for the first time that statin treatment benefits: women as well as men; the old as well as the young; those with low initial LDL levels as well as those with high initial levels; diabetics as well as nondiabetics. None of the earlier studies had been large enough to make this evident.

WOMEN

Women before the menopause have a much lower risk of coronary heart disease than men of the same age. However, after menopause their risk rises and, over a life span, coronary heart disease takes just as great a toll in women as in men. Nevertheless, there had been a tendency for physicians to regard women as "immune" and to undertreat their hypercholesterolemia. The statin studies have clearly shown that women benefit just as much from treatment as do men.

THE ELDERLY

Until recently physicians were somewhat reluctant to treat hypercholesterolemia in elderly patients. "Why bother them with yet another pill when they don't have much longer to live?" Only with the statin studies completed over the past few years has it become apparent that even patients over age

75 benefit from treatment – in relative terms – as much as younger people. Since the chances of a heart attack are much greater in the elderly, the absolute number of heart attacks prevented by treating 70-year old men is even greater than that prevented by treating men a decade or two younger.

In these days when life expectancy has risen to 75 in men and 80 in women, the number of good years of life conferred by treatment is large and the treatment is eminently worthwhile.

PATIENTS WITH NEAR-NORMAL LDL LEVELS

The question of "how high is high?" is a complicated one (53). The 2001 ATP III guidelines for physicians advised intensive treatment of very high risk patients (e.g. those with established coronary heart disease or with high risk due to diabetes), treatment designed to get their LDL down to 100 mg/dl (54). Some earlier studies had suggested that there was no significant reduction in risk for patients with an initial LDL level of 125 mg/dl or less (50). However, in the British Heart Protection Study, using simvistatin, subjects with initial LDL levels below 100 mg/dl, the "goal" by then-current standards, showed a significant further reduction of LDL levels *and* a further significant reduction of coronary heart disease risk with statin treatment (47). Some epidemiologic studies, particularly studies of the Chinese, had previously shown that coronary heart disease risk decreased with cholesterol levels even when the total cholesterol levels were in the 120–160 mg/dl range, i.e. LDL levels approximately 60–100 (55). However, the Heart Protection Study was the first to demonstrate directly that lowering LDL levels even below the 100 mg/dl level, previously considered to be ideal, does indeed confer additional benefit. The percentage reduction in risk was about the same as that in subjects with higher initial LDL levels. Very recently, using high doses of statins, it was shown that lowering LDL to a mean of 79 mg/dl arrested progression (measured by intravascular ultrasound), whereas lowering it only to a mean of 110 mg/dl still allowed further progression (56).

Overviews of the statin trials (38;51) show clearly that the lower the plasma LDL on treatment the lower the incidence of major end points (Figure 8.4). O'Keefe *et al.* plotted coronary heart disease events against LDL cholesterol levels in the available large primary prevention trials with statins (51). As seen in Figure 8.4, the number of events was directly proportional to LDL cholesterol level in both the control and treated groups but of course both were lower in those getting statins. Interestingly, extrapolation suggests that there might be no events at 55 mg/dl LDL cholesterol, but extrapolations often go away.

In any case the kinds of data discussed above will almost certainly bring pressure to reduce further the "goal LDL levels" recommended by the National Cholesterol Education Program.

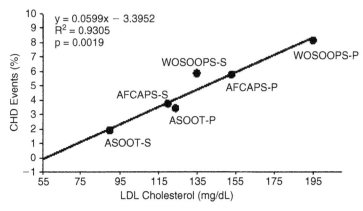

FIGURE 8.4   Pooled data from primary prevention statin trials, plotting the coronary heart disease event rates during the trial as a function of the LDL cholesterol levels during the trials (P, on placebo; S, on statin). The dramatic decrease in event rates as LDL falls is evident. Extrapolation of the data might suggest that the event rate would fall to zero at about 57 mg/dl LDL. Of course such extrapolations are not warranted but the national guidelines for LDL lowering are approaching just such a level. Source: reproduced from O'Keefe *et al.* (51) with permission from the American College of Cardiology Foundation.

DIABETICS

Most diabetic patients die of coronary heart disease, not coma or microvascular complications. For reasons still unclear, atherosclerosis proceeds at a higher rate in these patients and heart attacks occur about a decade earlier. When the diabetes is under good control the LDL levels are not necessarily elevated but a low HDL is the rule. So it was not certain that lowering LDL levels with statins would be effective. The Scandinavian Simvastatin Survival Study (57) and the British Heart Protection Study (47) provided the answer, and their results have been confirmed and extended in subsequent studies. Diabetics, whether with previous coronary heart disease or not, show as much benefit as nondiabetics.

WHAT CAN WE EXPECT IN THE FUTURE WITH THE STATINS?

Atherosclerosis is a disease of multiple etiologies. Proper clinical management should include intervention on all of them – dyslipidemia (high LDL; low HDL), cigarette smoking, hypertension, obesity, diabetes mellitus, lack of exercise. Yet intervention on just one risk factor – elevated LDL – has reduced coronary heart disease risk by 30–40 percent in the 5-year statin studies, indicating that hypercholesterolemia is a dominant determinant of clinical expression. What about the 60–70 percent of treated subjects who have a coronary event despite statin therapy? Several points can be made.

First, except for some very recent studies, the dose of statins used in clinical trials has been less than the maximum and less than optimal. Nor were adjuvant anti-lipid therapies included in an effort to obtain the maximal LDL lowering. Even this less-than-ideal intervention has reduced event rates dramatically. For primary prevention (Figure 8.4) the prediction from extrapolation is that with an on-treatment LDL level of about 57 mg/dl there might have been no events (51)! We recognize that extrapolations like this are not really justified and we have tongue firmly in cheek. Still, the data suggest that we may not yet have reached the limit of what can be achieved just by lowering LDL. With simultaneous attention to other causative factors the impact should be even greater.

Second, these studies have for the most part lasted for only 5 years. Percentage reduction in event rate might very well be significantly greater after 10 or 15 years of treatment.

Third, almost all the trials to date have been done in subjects with an average age of 50–60. We know that the arteries of these subjects harbor well-developed lesions even if they have no clinical manifestations of atherosclerosis. What, then, if intervention were started at age 40 or even 30, when the lesions were fewer and smaller? By how much would such early intervention further reduce the event rates? In individuals at unusually high risk, treatment should be started even earlier, even in childhood. A randomized, double-blind study of children with familial hypercholesterolemia ages 8–18 has demonstrated a significant slowing of intimal thickness in the carotid artery and no adverse effects on growth, hormone levels, sexual maturation, or liver function (58).

In short, the impact of the statins may ultimately exceed considerably that demonstrated in the clinical trials to date. However, if we hope to reach our goal of zero tolerance for myocardial infarction, we are probably going to have to start treatment earlier and also combine LDL lowering with equally vigorous attention to the other treatable risk factors. New modes of intervention under intensive current study include:

1. raising HDL or otherwise favoring reverse cholesterol transport (59);
2. inhibiting cholesterol absorption from the intestine (60;61);
3. attacking the inflammatory and immune processes contributing to the arterial lesion (62–64).

As a result of the statin studies the "ideal" LDL cholesterol level for subjects at high risk has dropped to about 70 mg/dl (65). Physicians are being urged to be more aggressive, using higher doses and using combination drug therapy.

Except for rhabdomyolysis, an extremely rare side effect, and an elevation of transaminase occasionally requiring discontinuation, the statins have proved to be remarkably safe, safer than aspirin (66). Some have proposed, only half in jest, that we put them "in the drinking water" as we do fluoride. In the United Kingdom, simvastatin (10 mg tablets) is already available over-the-counter; we in the United States are not ready to go quite that far (67). However, it is noteworthy that in this new Statin Era such proposals are no longer unthinkable.

NOTES

1 See R A. Fine's excellent history of this scandal (17).
2 Much of the following is based on an interview generously granted to me by Dr. Endo, on his own published accounts of the discovery (18;19), and on the several tributes to Endo in the Festschrift published in his honor, especially that by Brown and Goldstein (20).
3 Much of the following is based on interviews and personal communications from A.W. Alberts and P. Roy Vagelos, and on the book by Vagelos and Galambos, *Medicine, Science and Merck* (33).
4 The outstanding team at Merck that saw this project to completion included Doctors Georg Alber-Schonberg, Carl Hoffman, James MacDonald, Richard Monaghan, Arthur Patchett, and Ms Julie Chen.
5 Akira Yamamoto, personal communication.

REFERENCES

1. Steinberg, D. 1962. Chemotherapeutic approaches to the problem of hyperlipidemia. *Adv Pharmacol* 1:59–159.
2. Cottet, J., Redel, J., Krumm-Heller, C., and Tricaud, M.E. 1953. Hypocholesterolemic property of sodium phenylethylacetate (22 TH) in the rat. *Bull Acad Natl Med* 137:441–442.
3. Cottet, J., Mathivat, A., and Redel, J. 1954. Therapeutic study of a synthetic hypocholesterolemic agent: phenyl-ethyl-acetic acid. *Presse Med* 62:939–941.
4. Fredrickson, D.S. and Steinberg, D. 1956. Inhibitors of cholesterol biosynthesis and the problem of hypercholesterolemia. *Ann N Y Acad Sci* 64:579–589.
5. Steinberg, D., Fredrickson, D.S., and Avigan, J. 1958. Effects of 4-cholestenone in animals and in man. *Proc Soc Exp Biol Med* 97:784–790.
6. Steinberg, D. and Fredrickson, D.S. 1955. Inhibition of lipid synthesis by alpha-phenyl-N-butyrate and related compounds. *Proc Soc Exp Biol Med* 90:232–236.
7. Fredrickson, D.S. and Steinberg, D. 1957. Failure of alpha-phenylbutyrate and beta-phenylvalerate in treatment of hypercholesterolemia. *Circulation* 15:391–396.
8. Oliver, M.F. and Boyd, G.S. 1957. Effect of phenylethylacetic acid and its amide (hyposterol) on the circulating lipids and lipoproteins in man. *Lancet* 273:829–830.
9. Grande, F., Anderson, J.T., and Keys, A. 1957. Phenyl butyramide and the serum cholesterol concentration in man. *Metabolism* 6:154–160.
10. Tomkins, G.M., Sheppard, H., and Chaikoff, I.L. 1953. Cholesterol synthesis by liver. IV. Suppression by steroid administration. *J Biol Chem* 203:781–786.
11. Fredrickson, D.S., Peterson, R.E., and Steinberg, D. 1958. Inhibition of adrenocortical steroid secretion by delta 4-cholestenone. *Science* 127:704–705.

12. Tomkins, G.M., Nichols, C.W., Jr., Chapman, D.D., Hotta, S., and Chaikoff, I.L. 1957. Use of delta 4-cholestenone to reduce the level of serum cholesterol in man. *Science* 125:936–937.
13. Blohm, T.R. and MacKenzie, R.D. 1959. Specific inhibition of cholesterol biosynthesis by a synthetic compound (MER-29). *Arch Biochem Biophys* 85:245–249.
14. Avigan, J., Steinberg, D., Thompson, M.J., and Mosettig, E. 1960. Mechanism of action of MER-29, an inhibitor of cholesterol biosynthesis. *Biochem Biophys Res Commun* 2:63–65.
15. Steinberg, D., Avigan, J., and Feigelson, E.B. 1960. Identification of 24-dehydrocholesterol in the serum of patients treated with MER-29. *Prog Cardiovasc Dis* 2:586–592.
16. Avigan, J. and Steinberg, D. 1962. Deposition of desmosterol in the lesions of experimental atherosclerosis. *Lancet* 1:572.
17. Fine, R.A. 1972. *The Great Drug Deception: The Shocking Story of MER/29*. Stein and Day, New York.
18. Endo, A. 1988. Chemistry, biochemistry, and pharmacology of HMG-CoA reductase inhibitors. *Klin Wochenschr* 66:421–427.
19. Endo, A. 1992. The discovery and development of HMG-CoA reductase inhibitors. *J Lipid Res* 33:1569–1582.
20. Brown, M.S. and Goldstein, J.L. 2004. A tribute to Akira Endo, discoverer of a "penicillin" for cholesterol. *Atherosclerosis Suppl* 5:13–16.
21. Bentley, R. 2005. The development of penicillin: genesis of a famous antibiotic. *Perspect Biol Med* 48:444–452.
22. Endo, A., Kuroda, M., and Tsujita, Y. 1976. ML-236A, ML-236B, and ML-236C, new inhibitors of cholesterogenesis produced by Penicillium citrinium. *J Antibiot (Tokyo)* 29:1346–1348.
23. Endo, A., Kuroda, M., and Tanzawa, K. 1976. Competitive inhibition of 3-hydroxy-3-methylglutaryl coenzyme A reductase by ML-236A and ML 236B fungal metabolites, having hypocholesterolemic activity. *FEBS Letters* 72:323–326.
24. Endo, A., Tsujita, Y., Kuroda, M., and Tanzawa, K. 1979. Effects of ML-236B on cholesterol metabolism in mice and rats: lack of hypocholesterolemic activity in normal animals. *Biochim Biophys Acta* 575:266–276.
25. Tsujita, Y., Kuroda, M., Tanzawa, K., Kitano, N., and Endo, A. 1979. Hypolipidemic effects in dogs of ML-236B, a competitive inhibitor of 3-hydroxy-3-methylglutaryl coenzyme A reductase. *Atherosclerosis* 32:307–313.
26. Kuroda, M., Tsujita, Y., Tanzawa, K., and Endo, A. 1979. Hypolipidemic effects in monkeys of ML-236B, a competitive inhibitor of 3-hydroxy-3-methylglutaryl coenzyme A reductase. *Lipids* 14:585–589.
27. Brown, M.S., Faust, J.R., Goldstein, J.L., Kaneko, I., and Endo, A. 1978. Induction of 3-hydroxy-3-methylglutaryl coenzyme A reductase activity in human fibroblasts incubated with compactin (ML-236B), a competitive inhibitor of the reductase. *J Biol Chem* 253:1121–1128.
28. Chin, D.J., Luskey, K.L., Anderson, R.G., Faust, J.R., Goldstein, J.L., and Brown, M.S. 1982. Appearance of crystalloid endoplasmic reticulum in compactin-resistant Chinese hamster cells with a 500-fold increase in 3-hydroxy-3-methylglutaryl-coenzyme A reductase. *Proc Natl Acad Sci U.S.A.* 79:1185–1189.
29. Yamamoto, A., Sudo, H., and Endo, A. 1980. Therapeutic effects of ML-236B in primary hypercholesterolemia. *Atherosclerosis* 35:259–266.
30. Mabuchi, H., Haba, T., Tatami, R., Miyamoto, S., Sakai, Y., Wakasugi, T., Watanabe, A., Koizumi, J., and Takeda, R. 1981. Effect of an inhibitor of 3-hydroxy-3-methyglutaryl coenzyme A reductase on serum lipoproteins and ubiquinone-10-levels in patients with familial hypercholesterolemia. *N Engl J Med* 305:478–482.
31. Brown, A.G., Smale, T.C., King, T.J., Hasenkamp, R., and Thompson, R.H. 1976. Crystal and molecular structure of compactin, a new antifungal metabolite from Penicillium brevicompactum. *J Chem Soc (Perkin 1)*:1165–1170.

32. Fears, R., Richards, D.H., and Ferres, H. 1980. The effect of compactin, a potent inhibitor of 3-hydroxy-3-methylglutaryl coenzyme-A reductase activity, on cholesterogenesis and serum cholesterol levels in rats and chicks. *Atherosclerosis* 35:439–449.

33. Vagelos, P.R. and Galambos, L. 2004. *Medicine, Science and Merck*. Cambridge University Press, Cambridge:1–301.

34. Alberts, A.W., Chen, J., Kuron, G., Hunt, V., Huff, J., Hoffman, C., Rothrock, J., Lopez, M., Joshua, H., Harris, E., Patchett, A., Monaghan, R., Currie, S., Stapley, E., Albers-Schonberg, G., Hensens, O., Hirshfield, J., Hoogsteen, K., Liesch, J., and Springer, J. 1980. Mevinolin: a highly potent competitive inhibitor of hydroxymethylglutaryl-coenzyme A reductase and a cholesterol-lowering agent. *Proc Natl Acad Sci U.S.A.* 77:3957–3961.

35. 1984. The Lipid Research Clinics Coronary Primary Prevention Trial results. I. Reduction in incidence of coronary heart disease. *JAMA* 251:351–364.

36. 1984. The Lipid Research Clinics Coronary Primary Prevention Trial results. II. The relationship of reduction in incidence of coronary heart disease to cholesterol lowering. *JAMA* 251:365–374.

37. Scandinavian Simvistatin Survival Study Group. 1994. Randomised trial of cholesterol lowering in 4444 patients with coronary heart disease: the Scandinavian Simvistatin Survival Study (4S). *Lancet* 344:1383–1389.

38. Baigent, C., Keech, A., Kearney, P.M., Blackwell, L., Buck, G., Pollicino, C., Kirby, A., Sourjina, T., Peto, R., Collins, R., and Simes, J. 2005. Efficacy and safety of cholesterol-lowering treatment: prospective meta-analysis of data from 90,056 participants in 14 randomised trials of statins. *Lancet* 366:1267–1278.

39. Liao, J.K. 2005. Effects of statins on 3-hydroxy-3-methylglutaryl coenzyme a reductase inhibition beyond low-density lipoprotein cholesterol. *Am J Cardiol* 96:24F–33F.

40. LaRosa, J.C. 2001. Pleiotropic effects of statins and their clinical significance. *Am J Cardiol* 88:291–293.

41. Gould, A.L., Rossouw, J.E., Santanello, N.C., Heyse, J.F., and Furberg, C.D. 1998. Cholesterol reduction yields clinical benefit: impact of statin trials. *Circulation* 97:946–952.

42. Oliver, M.F. 1978. Cholesterol, coronaries, clofibrate and death. *N Engl J Med* 299:1360–1362.

43. Roberts, W.C. 1989. Safety of fenofibrate – US and worldwide experience. *Cardiology* 76:169–179.

44. Huttunen, J.K., Heinonen, O.P., Manninen, V., Koskinen, P., Hakulinen, T., Teppo, L., Manttari, M., and Frick, M.H. 1994. The Helsinki Heart Study: an 8.5-year safety and mortality follow-up. *J Intern Med* 235:31–39.

45. Brown, M.S. and Goldstein, J.L. 1986. A receptor-mediated pathway for cholesterol homeostasis. *Science* 232:34–47.

46. Mills, G.L. and Taylaur, C.E. 1971. The distribution and composition of serum lipoproteins in eighteen animals. *Comp Biochem Physiol B* 40:489–501.

47. 2002. MRC/BHF Heart Protection Study of cholesterol lowering with simvastatin in 20,536 high-risk individuals: a randomised placebo-controlled trial. *Lancet* 360:7–22.

48. Shepherd, J., Cobbe, S.M., Ford, I., Isles, C.G., Lorimer, A.R., Macfarlane, P.W., McKillop, J.H., and Packard, C.J. 1995. Prevention of coronary heart disease with pravastatin in men with hypercholesterolemia. West of Scotland Coronary Prevention Study Group. *N Engl J Med* 333:1301–1307.

49. Downs, J.R., Clearfield, M., Weis, S., Whitney, E., Shapiro, D.R., Beere, P.A., Langendorfer, A., Stein, E.A., Kruyer, W., and Gotto, A.M., Jr. 1998. Primary prevention of acute coronary events with lovastatin in men and women with average cholesterol levels: results of AFCAPS/TexCAPS. Air Force/Texas Coronary Atherosclerosis Prevention Study. *JAMA* 279:1615–1622.

50. Sacks, F.M., Pfeffer, M.A., Moye, L.A., Rouleau, J.L., Rutherford, J.D., Cole, T.G., Brown, L., Warnica, J.W., Arnold, J.M., Wun, C.C., Davis, B.R., and Braunwald, E. 1996. The effect of pravastatin on coronary events after myocardial infarction in patients with average cholesterol levels. Cholesterol and Recurrent Events Trial investigators. *N Engl J Med* 335:1001–1009.

51. O'Keefe, J.H., Jr., Cordain, L., Harris, W.H., Moe, R.M., and Vogel, R. 2004. Optimal low-density lipoprotein is 50 to 70 mg/dl: lower is better and physiologically normal. *J Am Coll Cardiol* 43:2142–2146.

52. Jacobs, D., Blackburn, H., Higgins, M., Reed, D., Iso, H., McMillan, G., Neaton, J., Nelson, J., Potter, J., Rifkind, B., Rossouw, J., Shekelle, R., and Yusuf, S. 1992. Report of the Conference on Low Blood Cholesterol: Mortality Associations. *Circulation* 86:1046–1060.

53. Steinberg, D. 1986. Plasma lipid levels: how much is too much? *Laboratory Management* 24:31–41.

54. 2001. Executive Summary of The Third Report of The National Cholesterol Education Program (NCEP) Expert Panel on Detection, Evaluation, And Treatment of High Blood Cholesterol In Adults (Adult Treatment Panel III). *JAMA* 285:2486–2497.

55. Chen, Z., Peto, R., Collins, R., MacMahon, S., Lu, J., and Li, W. 1991. Serum cholesterol concentration and coronary heart disease in population with low cholesterol concentrations. *Br Med J* 303:276–282.

56. Nissen, S.E., Tuzcu, E.M., Schoenhagen, P., Brown, B.G., Ganz, P., Vogel, R.A., Crowe, T., Howard, G., Cooper, C.J., Brodie, B., Grines, C.L., and DeMaria, A.N. 2004. Effect of intensive compared with moderate lipid-lowering therapy on progression of coronary atherosclerosis: a randomized controlled trial. *JAMA* 291:1071–1080.

57. Pyorala, K., Pedersen, T.R., Kjekshus, J., Faergeman, O., Olsson, A.G., and Thorgeirsson, G. 1997. Cholesterol lowering with simvastatin improves prognosis of diabetic patients with coronary heart disease. A subgroup analysis of the Scandinavian Simvastatin Survival Study (4S). *Diabetes Care* 20:614–620.

58. Wiegman, A., Hutten, B.A., de Groot, E., Rodenburg, J., Bakker, H.D., Buller, H.R., Sijbrands, E.J., and Kastelein, J.J. 2004. Efficacy and safety of statin therapy in children with familial hyperc-holesterolemia: a randomized controlled trial. *JAMA* 292:331–337.

59. Brewer, H.B., Jr. 2004. High-density lipoproteins: a new potential therapeutic target for the prevention of cardiovascular disease. *Arterioscler Thromb Vasc Biol* 24:387–391.

60. Feldman, T., Koren, M., Insull, W., Jr., McKenney, J., Schrott, H., Lewin, A., Shah, S., Sidisin, M., Cho, M., Kush, D., and Mitchel, Y. 2004. Treatment of high-risk patients with ezetimibe plus simvastatin co-administration versus simvastatin alone to attain National Cholesterol Education Program Adult Treatment Panel III low-density lipoprotein cholesterol goals. *Am J Cardiol* 93:1481–1486.

61. Davidson, M.H. 2003. Efficacy of simvastatin and ezetimibe in treating hypercholesterolemia. *J Am Coll Cardiol* 42:398–399.

62. Libby, P., Ridker, P.M., and Maseri, A. 2002. Inflammation and atherosclerosis. *Circulation* 105:1135–1143.

63. Glass, C.K. and Witztum, J.L. 2001. Atherosclerosis: the road ahead. *Cell* 104:503–516.

64. Hansson, G.K. 2001. Immune mechanisms in atherosclerosis. *Arterioscler Thromb Vasc Biol* 21:1876–1890.

65. Grundy, S.M., Cleeman, J.I., Merz, C.N., Brewer, H.B., Jr., Clark, L.T., Hunninghake, D.B., Pasternak, R.C., Smith, S.C., Jr., and Stone, N.J. 2004. Implications of recent clinical trials for the National Cholesterol Education Program Adult Treatment Panel III guidelines. *Circulation* 110:227–239.

66. Tobert, J.A. 1996. Statin therapy and CHD. *Lancet* 347:128.

67. Gotto, A.M., Jr. 2005. Over-the-counter statins and cardiovascular disease prevention: per-spectives, challenges, and opportunities. *Clin Pharmacol Ther* 78:213–217.

CHAPTER 9

Summing up

THE LONG, UPHILL ROAD FOR THE LIPID HYPOTHESIS

It was a bumpy ride but we made it. No one any longer doubts the wisdom of lowering blood cholesterol. Extrapolating from the exciting results of the 5-year statin studies, we can safely predict that when treatment is started earlier and continued for a longer time, heart attack rates will drop even more strikingly. Hopefully, early and intensive medical attention to hypercholesterolemia along with equally intensive attention to the other major risk factors will eventually reduce sharply the need for interventional cardiology. Some of the key milestones leading to the ultimate acceptance of the lipid hypothesis are summarized in Table 9.1.

WHY DID THE CONTROVERSY LAST SO LONG?

From this analysis of the controversy, and a previously published analysis (1), there emerge at least ten reasons why the proposal to treat hypercholesterolemia was so strongly resisted for so long. Those reasons include the following.

1. Dismissal of Anitschkow's rabbit model and other animal models as not relevant to the human disease.

2. Misguided search for a single cause in a complex disease of multiple etiology. For example, the argument went: "If only a subset of cases show hypercholesterolemia then hypercholesterolemia cannot be a major causative factor." Closely related to this misconception is the next point.

3. Confusion as to what constitutes a "normal" blood cholesterol level: unwillingness to accept the notion that a very large fraction of our nominally "healthy" population actually has an unhealthily high cholesterol level and requires treatment.

4. Undue focus on the advanced, complex lesions rather than the initiating factors.

The Cholesterol Wars by D. Steinberg.
Copyright © 2007 Elsevier Inc. All rights of reproduction in any form reserved.

TABLE 9.1   Milestones on the road to acceptance of the lipid hypothesis

1913	Anitschkow – The cholesterol-fed rabbit develops high blood cholesterol levels and arterial lesions similar to those seen in human atherosclerosis.
1939	Müller – Familial hypercholesterolemia: xanthomatosis and angina pectoris both linked to hypercholesterolemia.
1949	Gofman – the lipoproteins of human plasma and their correlation with the risk of coronary heart disease.
1952–54	Kinsell and Ahrens – blood cholesterol in normal subjects is elevated by saturated fats in the diet.
1957	Keys' Seven Country Study – incidence of coronary heart disease directly correlated to hypercholesterolemia and to dietary fat intake in an international cohort study.
1961	Framingham Study – within a typical American community, coronary heart disease risk is highest in groups with highest blood cholesterol levels.
1961	American Heart Association – first endorsement of the "prudent" low-fat diet.
1964	Nobel Prize to Konrad Bloch for elucidating the pathway for cholesterol biosynthesis, paving the way for the statins a decade later.
1966–69	Leren, Miettinen, Dayton, and others – reducing blood cholesterol levels by reducing saturated fat intake reduces coronary heart disease risk. However, data are not considered persuasive.
1974	Goldstein and Brown – identification of the LDL receptor as the key regulator of cholesterol and lipoprotein metabolism.
1976	Endo – discovery of the first statin drug (compactin), a powerful inhibitor of HMGCoA reductase, effective in lowering blood cholesterol in animals and humans.
1980	Merck – discovery of mevinolin (lovastatin), later to become the first statin to reach the market.
1984	Lipid Research Clinics' Coronary Primary Prevention Trial shows significant reduction in coronary heart disease primary events in hypercholesterolemic men treated with cholestyramine.
1984	NIH Consensus Conference on Lowering Blood Cholesterol to Prevent Heart Disease – NIH, for the first time, declares lowering blood cholesterol to be a national public health goal.
1984	NIH coordinates a national program for teaching physicians and patients how to diagnose and deal with dyslipidemia: the National Cholesterol Education Program.
1985	Nobel Prize to Brown and Goldstein for groundbreaking work on regulation of cholesterol and LDL metabolism.
1994	The Statin Era – Scandinavian Simvastatin Survival Study (4S), first truly large-scale, randomized, double-blinded trial to show that aggressive treatment (with simvastatin) not only reduces coronary heart disease mortality but also decreases all-cause mortality.
1995–present	A stunning series of studies demonstrating the power and the safety of statin intervention.

5. Confusion between cholesterol in the diet and cholesterol in the blood. Of course it's the latter that counts; the diet is relevant but mainly as one of many other determinants of blood cholesterol level.

6. Reluctance of practitioners in the 1970s and 1980s to grapple with the (to them) still elusive plasma lipoproteins and their complex metabolism.

7. The limited and relatively unsatisfactory dietary and drug regimens available for controlling hypercholesterolemia prior to the statin era ("What's the difference; we can't do much about hypercholesterolemia anyway").

8. The absence, until fairly recently, of a consensus on the mechanisms linking cholesterol and lipoproteins to the damage in the artery wall.

9. Preoccupation of the cardiologists with their new and exciting diagnostic and interventional tools; impatience with the notion of preventive cardiology.

10. Most important of all, resistance to the need to synthesize evidence of several different kinds – epidemiologic evidence, experimental observations in animals, genetic evidence, clinical observations, and clinical trial data – in evaluating the true strength of the lipid hypothesis. The early clinical trial results, while weaker than might have been desired, were nevertheless impressive if they were weighed *in the context of all the other available lines of evidence*.

SOME THOUGHTS ON HYPOTHESIS TESTING

A major message from the history presented here is item 10 in the list above, namely: *cumulative evidence* and *evidence of different kinds* must be taken into account in evaluating postulated causal relations and certainly must be taken into account when deciding what to do about them. The lipid hypothesis proposed that hypercholesterolemia was a causative factor in human atherosclerosis. It did not propose that hypercholesterolemia was the only cause but that it was at least a quantitatively significant factor. An implied corollary was that appropriate intervention to correct hypercholesterolemia might reduce the rate of progression of atherosclerosis and its clinical manifestations. To many researchers and clinicians, the *only definitive test* of the lipid hypothesis was going to be the clinical trial: the gold standard single variable, randomized, double-blind, placebo-controlled intervention trial. And that indeed is the crucial test of the corollary. However, results of any clinical trial need to be evaluated in the light of prior information bearing on the likelihood of the hypothesis. For example, let us say that a clinical trial of cholesterol lowering yields a 20 percent decrease in event rate with a p value of 0.07. To a skeptic (or to one unfamiliar with the many other lines of evidence supporting the lipid hypothesis) such a result ($p > 0.05$) might be the death knell of the hypothesis. On the other hand, to one familiar with the extensive ancillary evidence supporting the hypothesis, the same result would probably be regarded as importantly supportive. It would at

the very least lead to additional studies and might even lead to recommendations that treatment of high-risk patients be considered.[1]

The notion of weighing all the findings bearing on a hypothesis rather than
just looking at the clinical trial result under evaluation in isolation seems self-
evident. It was actually formally put forward over 200 years ago by Bayes and
has been extended and formalized in a number of ways over the years (4;5).
Basially, Bayes said, "From looking at all the relevant information before we carried out this trial, what was the likelihood going in that the hypothesis was
valid?" That would lead to what he called a *prior probability* and could considerably affect the evaluation of the results of the current trial and the decision to
act on the basis of its results. When all of the data under consideration are of a
single kind, calculating the prior probability is fairly straightforward. Indeed,
Bayesian statistics is now being used more commonly for analysis of such data.
When, however, the prior evidence includes qualitatively distinct observations,
it is not clear how to weigh and take them into account. Unfortunately, we still
do not have a consensus on the Bayesian approach nor the instruments for formally quantifying the weights of different lines of evidence for inclusion in a
probability algorithm. Much as we would like to avoid subjective weightings in
the evaluation of biomedical hypotheses, in order to reach sound judgments as
to public health policy we simply have to use all of the evidence available. The
cholesterol controversy could have been resolved much earlier if *all* of us had
looked at *all* of the evidence.

HAVE WE GONE AS FAR AS WE CAN GO WITH
LDL LOWERING?

It is too early to say to what extent aggressive cholesterol lowering will ultimately reduce morbidity and mortality from coronary heart disease. The statin
trials to date have reduced event rates by 20–30 percent. This corresponds to
the prevention of over 100,000 coronary events per year! We can all take pride
in this remarkable achievement. However, looked at another way, fully 70–80
percent of the participants in these trials went on to die or have a myocardial
infarction despite statin treatment. We are by no means home free. On the
other hand, these trials were carried out in middle-aged populations. Wouldn't
the salvage rate be higher if we started treating earlier?

It is well established that atherosclerosis begins in childhood and that the
same risk factors that predict susceptibility in adults also predict risk in children (6;7). The seeds are sown early. Starting treatment at an earlier age, *before*
lesions begin to express themselves clinically, therefore makes logical sense.
We already treat very high risk patients (e.g. patients with familial hypercholesterolemia) in childhood and there is evidence that it works and that it is

safe. In children aged 8–18, statin treatment lowered LDL levels by 24 percent and significantly slowed the rate of progression of intima-medial thickening of the carotid artery. There were no adverse effects on growth, hormone levels, or sexual maturation (8). However, until long-term studies have been done we can't say with certainty how much better the results will be as a result of starting earlier. Still, some exciting new results suggest that we may be pleasantly surprised.

A new gene importantly involved in regulation of the LDL receptor, PCSK9, has recently been identified (9;10). The gene codes for a secreted protease that acts to decrease the number of LDL receptors expressed in the liver. Over-expression of the gene (or gain-of-function mutations) lowers LDL receptor number and thus raises LDL levels; knocking out the gene (or loss-of-function mutation) increases LDL receptor number and thus lowers LDL levels. PCSK9 does not regulate at the transcriptional or translational level but rather in some fashion suppresses the level of expression of the LDL receptor at the surface of the hepatocyte, although the precise molecular mechanism is still uncertain.

The key point of interest here is that Cohen et al. in Hobbs' laboratory (11;12) have studied coronary heart disease risk in subjects with nonsense mutations in PCSK9 that cause low LDL levels, presumably from birth. The LDL level in these subjects was reduced by 28 percent. *The coronary heart disease risk was reduced by fully 88 percent!* Now in the 5-year statin trials the drop in LDL was also in the neighborhood of 20–30 percent but the decrease in risk was only about 30 percent, not 88 percent. The implication is that having a low LDL from birth almost triples the magnitude of the effect on risk compared to the risk reduction found in a 5-year trial of middle-aged people. The numbers of subjects studied by Hobbs and others so far have been small and further studies are needed. If the data hold up, it may be that lowering LDL from an early age could more than double the salvage rate (13). Drugs that regulate the expression and function of PCSK9 may soon join the ranks of cholesterol-lowering agents. Working by a different mechanism, the effects of such agents may be additive to the effects of statins and ezetimibe.

STATINS FOR EVERYONE?

The statins are remarkably safe – probably safer than aspirin. In patients at even moderately high risk of heart attack the potential benefits outweigh the relatively minor risk. However the statins do have some side effects, albeit rare and usually minor. So rare are these side effects that making statins an over-the-counter drug has been proposed. It has even been proposed, only half facetiously, that we may some day put statins in the drinking water, as we do with fluoride; or in salt, as we do with iodide; or in multivitamin capsules. The rationale behind these startling proposals is that about half the people who die every year die because of atherosclerosis and its complications – heart attacks and strokes. Many more have nonfatal heart attacks that might be prevented. It is perfectly legitimate to ask: "If almost everyone is at risk, why not treat almost everyone?" Personally, I don't think we are quite ready for such futuristic scenarios. If even safer statins should come along and we have a drug with absolutely no side effects – zero – then this question can be revisited.

NEW PREVENTIVE MEASURES ON THE HORIZON

Treatment of hypercholesterolemia is already remarkably effective but even this will predictably improve with time. For example, the recently introduced inhibitor of cholesterol absorption from the intestine (ezetimibe) has significantly enhanced the effectiveness of statins. Other targets, such as the process by which the lipoproteins are fabricated in the liver, are under investigation.

Low HDL is just as good a predictor of coronary heart disease risk as high LDL. Until recently the only available drug that raised HDL was nicotinic acid and indeed the effectiveness of nicotinic acid in reducing risk may depend in part on its HDL-raising effect (14). A new and highly effective HDL-raising drug, torcetrapib, was introduced in 2003, and is well along in clinical testing (15;16). The drug works by inhibiting the cholesterol ester transferase protein (CETP). It can increase HDL levels by as much as 50–60 percent. Since a 1 percent increase in HDL correlates with a 2–3 percent decrease in risk, there is understandably a high level of interest in this and other approaches to raising HDL (17). Unfortunately, the clinical trial of torcetrapib was aborted just as this book was going to press; there were more cardiovascular events in the treated group than in the placebo group. The mechanism underlying the toxicity is being intensively explored.

The response of the artery wall exposed to high concentrations of LDL begins with the entry into the artery wall of circulating white blood cells (monocytes and lymphocytes) and includes many features in common with inflammatory processes elsewhere in the body (18–21). The invading cells interact with the cells of the artery wall and with each other in complex ways, partly through the secretion of substances that influence the growth of cells, the laying down of fibrous elements, blood clotting, and so on. Drugs targeted at modulating this complex inflammatory process promise to provide a new category of agents acting directly at the arterial wall. Such agents should be additive or even synergistic with cholesterol-lowering drugs and many drug companies are exploring new drugs in this category.

A BOLD PROPOSAL FOR A LONG-TERM APPROACH TO CONTROLLING HYPERCHOLESTEROLEMIA IN THE UNITED STATES AND OTHER DEVELOPED COUNTRIES

The use of cholesterol-lowering drugs is slowing the epidemic of coronary heart disease to a walk, and for that we are all grateful. It may very well slow down even further in the coming decade. But do we want to be a nation in which a large fraction of the population has to pop two, three or four pills every day to ward off coronary heart disease? People with genetically determined forms of dyslipidemia will continue to require drug treatment but for the bulk of the general population there is an alternative although it would be challenging in the extreme.

Over a period of decades, and it certainly would take many decades, hopefully we could gradually wean our population to a diet and exercise regimen that would simulate that of the Japanese in 1949. At that time the death rate from coronary heart disease in Japan was one-tenth that in the United States (22). Note that that was despite a higher prevalence of high blood pressure and cigarette smoking in Japan than in the United States. These are, of course, two of the most significant risk factors for coronary heart disease. The heart attack rate in Japan is now rising, thanks to their increasing dietary intake of saturated fat and a rapidly increasing incidence of obesity.

Would such a revolutionary plan work in the United States and other Western societies? Or are the Japanese protected on some genetic basis? The studies of Japanese who migrated to San Francisco suggest not; when they migrated, their blood cholesterol rose and their heart attack rate rose with it. The major reason for the low rate in native Japanese in the 1940s was environmental, i.e. dietary habits and exercise habits.

China, like Japan, has also enjoyed a very low heart attack rate and, like Japan, has had a very low average blood cholesterol level. In 2002, Joseph L. Witztum and I were invited to visit several medical schools in China and to lecture on atherosclerosis and coronary heart disease. In preparation for those lectures I read whatever I could find about coronary heart disease in modern China. From the available literature it was clear that coronary heart disease was still a minor disease problem in China, as it is in Japan. But it was equally clear that it was increasing rapidly and was going to go on increasing if nothing was done.

As we entered the main square in Cheng-Du we saw the imposing larger-than-life statue of Chairman Mao on the left and a McDonald's Golden Arches on the right! I later learned that there are already more than 4,000 McDonald's franchises in China and Colonel Sanders is a close second.

Also there were already a lot of upscale restaurants where beef and pork and creamy sauces and butter and cheese are to be had. The once Spartan, low-fat diet of the Chinese is being rapidly Westernized. Rises in blood cholesterol have already been documented, especially in the cities. We were expecting to see, as we did, crowds of bicycles on the streets but were not really prepared to see the huge numbers of automobiles. You could tell that this was a relatively recent phenomenon by the number of bad drivers – really bad drivers. Cars are more affordable (though still pretty expensive) and those who can afford them are happily exchanging their two-wheelers for four-wheelers.

The combination of calories being more readily available (and available in appetizing forms), and less need for exercise is a sure recipe for a rapid increase in the modern plague of the Western world – obesity. The increase in obesity has also been documented in China. With it comes an inevitable increase in high blood pressure and diabetes – and a greater susceptibility to coronary disease.

Our message to the Chinese was that they still had the opportunity to fore-stall an epidemic of coronary heart disease. If blood cholesterol levels continue to increase in China and if obesity becomes more prevalent, the Chinese could catch up to us in terms of coronary heart disease death rates in just a few decades. Many Chinese physicians are fully aware of these basic facts about the situation. Will they be able to do anything about it? My guess is probably not, at least not right away. The economic boom in China is truly amazing and at least the rich and powerful are riding high. The appeal of American comfort foods is all but irresistible and so is the charisma of the car. Only an edict from the Central Committee, if even that, is going to stem the tide of Westernization driven by private enterprise.

To be sure there may be additional factors that confer on the Japanese and the Chinese their wonderfully low heart attack rate, but it appears that diet is the main reason. If so, we could reduce heart attack rates in the United States by almost 90 percent by radically modifying our diet, "Nipponizing" it, if you will. Such a proposal would, understandably, be met with loud guffaws.

Actually, just such a proposal was made almost 40 years ago by a farsighted pioneer in nutrition, William E. Connor. He urged the American Heart Association to recommend what he called an "alternative diet," a diet that he had shown lowered blood cholesterol in hyperlipidemic subjects by 26 percent (23). However, he was overruled by more conservative voices.

In the 1960s, the Surgeon General's proposal that everyone stop smoking was also met with loud guffaws but look at what has happened. Public education and social pressure, helped along by a few statutes, have radically altered the way smoking is regarded. Slowly but surely the number of cigarette smokers is decreasing. Smoking is addictive but saturated fat is not (at least not to the same extent). Therefore, I think it should be less difficult to kick the saturated fat habit than to kick the smoking habit. Not that it would be a breeze.

Of course not. It would require national public education programs, cooperation from the food manufacturers and purveyors, and constant reinforcement from health professionals. And it would not happen over decades but over several generations. However, as Chairman Mao said, "The longest journey begins with one step."

NOTE

1 Space does not allow a discussion of the widespread misunderstanding of the true meaning of the *p* value and its limitations; the interested reader will find clear expositions in the papers by Peto *et al.* (2) and by Goodman (3).

REFERENCES

1. Steinberg, D. 1989. The cholesterol controversy is over. Why did it take so long? *Circulation* 80:1070–1078.
2. Peto, R., Pike, M.C., Armitage, P., Breslow, N.E., Cox, D.R., Howard, S.V., Mantel, N., McPherson, K., Peto, J., and Smith, P.G. 1976. Design and analysis of randomized clinical trials requiring prolonged observation of each patient. I. Introduction and design. *Br J Cancer* 34:585–612.
3. Goodman, S.N. 1999. Toward evidence-based medical statistics. 1: The P value fallacy. *Ann Intern Med* 130:995–1004.
4. Cornfield, J. 1969. The Bayesian outlook and its application. *Biometrics* 25:617–657.
5. Goodman, S.N. 1999. Toward evidence-based medical statistics. 2: The Bayes factor. *Ann Intern Med* 130:1005–1013.
6. Berenson, G.S., Srinivasan, S.R., Bao, W., Newman, W.P., III, Tracy, R.E., and Wattigney, W.A. 1998. Association between multiple cardiovascular risk factors and atherosclerosis in children and young adults. The Bogalusa Heart Study. *N Engl J Med* 338:1650–1656.
7. McMahan, C.A., McGill, H.C., Gidding, S.S., Malcom, G.T., Newman, W.P., Tracy, R.E., and Strong, J.P. 2007. PDAY risk score predicts advanced coronary artery atherosclerosis in middle-aged persons as well as youth. *Atherosclerosis* 190:370–377.
8. Wiegman, A., Hutten, B.A., de Groot, E., Rodenburg, J., Bakker, H.D., Buller, H.R., Sijbrands, E.J., and Kastelein, J.J. 2004. Efficacy and safety of statin therapy in children with familial hypercholesterolemia: a randomized controlled trial. *JAMA* 292:331–337.
9. Abifadel, M., Varret, M., Rabes, J.P., Allard, D., Ouguerram, K., Devillers, M., Cruaud, C., Benjannet, S., Wickham, L., Erlich, D., Derre, A., Villeger, L., Farnier, M., Beucler, I., Brucker, T.E., Chambaz, J., Chanu, B., Lecerf, J-M., Luc, G., Moulin, P., Weissenbach, J., Prat, A., Krempf, M., Junien, C., Seidah, N., and Beileau, C. 2003. Mutations in PCSK9 cause autosomal dominant hypercholesterolemia. *Nat Genet* 34:154–156.
10. Maxwell, K.N., Soccio, R.E., Duncan, E.M., Sehayek, E., and Breslow, J.L. 2003. Novel putative SREBP and LXR target genes identified by microarray analysis in liver of cholesterol-fed mice. *J Lipid Res* 44:2109–2119.
11. Cohen, J., Pertsemlidis, A., Kotowski, I.K., Graham, R., Garcia, C.K., and Hobbs, H.H. 2005. Low LDL cholesterol in individuals of African descent resulting from frequent nonsense mutations in PCSK9. *Nat Genet* 37:161–165.

12. Cohen, J.C., Boerwinkle, E., Mosley, T.H., Jr., and Hobbs, H.H. 2006. Sequence variations in PCSK9, low LDL, and protection against coronary heart disease. *N Engl J Med* 354:1264–1272.

13. Brown, M.S. and Goldstein, J.L. 2006. Biomedicine. Lowering LDL – not only how low, but how long? *Science* 311:1721–1723.

14. Rubins, H.B., Robins, S.J., Collins, D., Fye, C.L., Anderson, J.W., Elam, M.B., Faas, F.H., Linares, E., Schaefer, E.J., Schectman, G., Wilt, T.J., and Wittes, J. 1999. Gemfibrozil for the secondary prevention of coronary heart disease in men with low levels of high-density lipoprotein cholesterol. Veterans Affairs High-Density Lipoprotein Cholesterol Intervention Trial Study Group. *N Engl J Med* 341:410–418.

15. Brousseau, M.E., Schaefer, E.J., Wolfe, M.L., Bloedon, L.T., Digenio, A.G., Clark, R.W., Mancuso, J.P., and Rader, D.J. 2004. Effects of an inhibitor of cholesteryl ester transfer protein on HDL cholesterol. *N Engl J Med* 350:1505–1515.

16. Clark, R.W., Sutfin, T.A., Ruggeri, R.B., Willauer, A.T., Sugarman, E.D., Magnus-Aryitey, G., Cosgrove, P.G., Sand, T.M., Wester, R.T., Williams, J.A, Perlman, M., and Bamberger, M.J. 2004. Raising high-density lipoprotein in humans through inhibition of cholesteryl ester transfer protein: an initial multidose study of torcetrapib. *Arterioscler Thromb Vasc Biol* 24:490–497.

17. Brewer, H.B., Jr. 2004. Increasing HDL Cholesterol Levels. *N Engl J Med* 350:1491–1494.

18. Glass, C.K. and Witztum, J.L. 2001. Atherosclerosis: the road ahead. *Cell* 104:503–516.

19. Ross, R. 1999. Atherosclerosis – an inflammatory disease. *N Eng J Med* 340:115–126.

20. Libby, P., Hansson, G.K., and Pober, J.S. 1999. Atherogenesis and inflamation. In *Molecular Basis of Cardiovascular Disease*, K.R. Chien, ed. Philadelphia, W.B.Saunders: 349–366.

21. Hansson, G.K. 2001. Immune mechanisms in atherosclerosis. *Arterioscler Thromb Vasc Biol* 21:1876–1890.

22. Kimura, N. 1956. Analysis of 10,000 postmortem examinations in Japan. In *World Trends in Cardiology: Selected Papers from Second World Congress of Cardiology*, A. Keys and P.D. White, eds. Hoeber-Harper, New York:22–33.

23. Connor, W.E. 1968. Measures to reduce the serum lipid levels in coronary heart disease. *Med Clin North Am* 52:1249–1260.

Frequently cited criticisms

In the course of presenting this history of how the many lines of evidence supporting the "lipid hypothesis" were developed, I have dealt with many or most of the arguments presented over the years by opponents. Any remaining doubts about the correctness of the lipid hypothesis were swept away by the large-scale statin trials (see Chapter 8). However, for historical reasons, it may be useful to analyze here some of the most frequently cited criticisms. In each of the responses reference is made to the chapters in the text that deal with the criticism more extensively.

"LOWERING CHOLESTEROL LEVELS WITH DRUGS MAY BE DANGEROUS"

This concern was originally based in part on the failure of the early intervention trials to show a statistically significant decrease in all-cause mortality even though they did in fact decrease coronary mortality rates. That was the case in the Coronary Primary Prevention Trial using cholestyramine (1;2) and in the Helsinki trial using gemfibrozil (3). The argument made was: "What's the point of preventing cardiac deaths if you just substitute deaths from other causes?"

In both the studies cited above there were more deaths due to trauma (lumping together automobile accidents, suicide, homicide) in the treated group than in the placebo group but the numbers were very small and the differences were not statistically significant. For the cohort as a whole there was *neither a significant increase nor a significant decrease in all-cause mortality.* The key point is that the number of subjects studied was too small to have yielded a statistically significant result on the issue of all-cause mortality even if there had been a difference. This was especially true in primary prevention studies where the event rates are lower. Also, in these early studies the blood cholesterol levels were only lowered by about 10–15 percent. As pointed out by Collins *et al.* (4), the number of subjects randomized would have had to be considerably larger or the drop in cholesterol level considerably greater in order to have had any chance of detecting a significant decrease in all-cause mortality (see pp. 154–155).

The Cholesterol Wars by D. Steinberg.
Copyright © 2007 Elsevier Inc. All rights of reproduction in any form reserved.

Wysowski and Gross (5) tracked down the details of the "trauma deaths" in these two studies and found that attribution of these deaths to cholesterol lowering per se made absolutely no sense (see p. 155).

Clofibrate, used in a series of early drug intervention studies, significantly decreased the incidence of coronary events, again confirming the basic lipid hypothesis (6;7). However, clofibrate turned out to be toxic (8). There was actually a significantly greater incidence of deaths related to the gastrointestinal tract in the treated group. However, there was no evidence that this was caused by the lowering of the cholesterol levels per se. A close structural relative of clofibrate, gemfibrozil, was used in a later trial. This analogue lowered cholesterol levels to the same extent but did not affect all-cause mortality nor the incidence of gastrointestinal disorders (3). The culprit was not the lowering of cholesterol levels, but the toxicity of the clofibrate molecule (see pp. 131–134).

Finally, thanks to the statin trials, with a greater drop in cholesterol levels and a much larger number of subjects in each study, there has been a statistically significant drop in *both coronary heart disease mortality and all-cause mortality*. In these trials there was a 20–30 percent decrease in fatal or nonfatal myocardial infarction and a similar decrease in all-cause mortality. There was no increase in any individual category of deaths, including cancer and trauma (9–11) (see pp. 184–188).

"HAVING AN INNATELY LOW CHOLESTEROL LEVEL IS DANGEROUS"

Before the statin trial results were available, the concerns about the safety of lowering blood cholesterol were taken seriously. Indeed, in 1992, the National Heart Lung and Blood Institute organized a Conference on Low Blood Cholesterol: Mortality Associations (12). The participants reviewed data from 20 cohort studies that had reported associations of low blood cholesterol with increases in frequency of noncardiovascular diseases. Some of these were attributable to the fact that some diseases, including cancer, chronic pulmonary disease, and cirrhosis, cause lowering of cholesterol levels even at early, subclinical stages. In other words the subjects with the lowest levels of blood cholesterol included these individuals who were already sick and at greater risk for death over the ensuing years, not because their cholesterol levels were low, but because they already had an illness that caused their cholesterol level to be low. Not all of the data could be explained this way but after extended discussion the conclusion of the participants was that "Definitive interpretation of the associations observed was not possible, although *most of the participants considered it likely that many of the statistical associations … are explainable by confounding in one form or another."*

However, the report of this Conference was accompanied by an Editorial by Hulley *et al.* (13) that interpreted the available evidence very differently. They said that "it now seems unwise to treat high blood cholesterol with drugs" and "we should draw back from universal screening and treatment of blood cholesterol for primary prevention in adults as well" and, excepting women with existing coronary heart disease or at comparably high risk of coronary heart disease death, "it no longer seems wise to treat high blood cholesterol in women." Needless to say this dramatic editorial garnered more attention in the press than the moderate, "we-don't really-have-the-answers" NHLBI Conference report itself. Fortunately, there were not many scientists who lined up with Hulley. A workshop held the following year in Milan under the auspices of the International Society and Federation of Cardiology and the International Task Force for the Prevention of Coronary Heart Disease pointed out the errors in the Hulley Editorial (14).

One obvious fallacy in the thinking of Hulley and the others who were afraid that treatment of hypercholesterolemia might be harmful is that the levels of blood cholesterol reached on drug treatment are still far above the levels below which the nadir of the "J" curve lies. In other words, *even if* there were an increased risk associated with a low cholesterol level it would only begin at 160 mg/dl or lower. Even with aggressive measures to lower blood cholesterol, rarely if ever did the levels go that low.

There is no evidence that the apparent increase in mortality associated with having a low cholesterol level reflects a causal relationship. Rather the evidence is that the low cholesterol levels are the *result* of these conditions, not the cause. One of the best studies to establish the correctness of this interpretation is that of Iribarren *et al.* (15). They studied a cohort of almost 6,000 Japanese-American men aged 45–68 years without prior history of cardiovascular disease over a 16-year period from 1973 to 1988. During this time blood cholesterol levels were measured at baseline and at two intervals thereafter. This allowed them to compare mortality in those whose cholesterol level fell over this time interval with the mortality in those who had a low level initially but in whom that low level was stable. They found that in those whose cholesterol levels fell (spontaneously) there was an accompanying significant increase in the frequency of other diseases including nonmalignant liver disease, total cancer, and, most noticeably, cancers of the esophagus, prostate, and hemopoietic systems. In contrast, those who had a low but stable blood cholesterol level showed no increased mortality risk although there was some association of very low cholesterol levels with fatal hemorrhagic stroke.

In his Introduction to the Workshop on Low Blood Cholesterol: Health Implications, Barry Lewis wrote (14):

> If lowered or innately low blood cholesterol levels confer health hazards, this must profoundly influence orthodox views on coronary heart disease prevention strategies. Conversely, if cholesterol lowering does little or no harm, then we are in

danger of withholding or under-utilizing an effective component of risk reduction
strategy as a result of inappropriate but widely expressed concerns.

At the conclusion of the Workshop he added: "The consistent answer from the
presentations is that the second of these is true."

Currently more and more physicians are lowering LDL more and more
aggressively. Is there a danger that there will be ill effects? Probably not, for the
following reasons. There are kindreds with genetically determined hypobetalipoproteinemia, some with total cholesterol levels of less than 50 mg/dl and
LDL levels below 10 or 15 mg/dl. Yet the members of these kindreds not only
fail to show any ill effects, they actually have increased longevity (16). In the
animal kingdom there are many species with LDL levels below 50 mg/dl (mice,
rats, dogs) and, at birth, humans have similarly low levels. These low levels are
sufficient to provide all the cholesterol the cells need. The LDL receptor pathway insures that that is the case. Any drop in LDL levels is immediately com-
pensated for by an increase in LDL receptor number and an increase in cellular
LDL uptake. The cell content of cholesterol is jealously guarded. In the recent
statin trials, LDL has been lowered to 70 mg/dL or less with no evident toxicity
and with greater reeduction in risk (see pp. 185–186 and Figure 8.4).

"DIETARY FAT INTAKE IS IRRELEVANT TO
CORONARY HEART DISEASE RISK"

As discussed in detail in Chapter 3, pp. 41–46, there is solid evidence, evidence
from controlled metabolic ward trials that substituting polyunsaturated or
monounsaturated fat for an equicaloric amount of saturated fat reduces blood
cholesterol levels. These are rigorously designed trials in which each individual serves as his or her own control (17–19). The studies show that the magnitude of the response differs from person to person but the drop in total
cholesterol can be as much as 15–25 percent. From the epidemiological data
relating cholesterol level to coronary heart disease risk, such a drop would be
predicted to decrease risk by as much as 30–50 percent.

The criticism that these results from metabolic ward studies cannot be
extrapolated to the general population is best answered by referring to the
huge Diet–Heart Feasibility trial (20). About 1,000 men aged 45–54 were randomly selected from census tracts and studied for a year on a diet high in saturated fat or a diet high in polyunsaturated fat using a double-blind design. The
subjects were free-living and got their fat-rich foods from a central depot
(coded to maintain the double-blind). The men on the polyunsaturated fat-
rich diet decreased their blood cholesterol level by 11 percent.

The response was quite variable from individual to individual. About 15
percent of the men dropped their cholesterol level by more than 29 percent,

and 27 percent of the men dropped their level by less than 5 percent (see Table 3.1). This wide variability in diet responsiveness may be why epidemiologic studies within a given population often fail to show a correlation between dietary fat intake and cholesterol levels, as discussed below. The definitive Diet–Heart Study was never undertaken because it would have been much too expensive. The fact that the feasibility study documented the effect of saturated fat on cholesterol levels in 1,000 men has often been lost sight of.

It is not true that the diet-heart hypothesis has never been tested directly. There have been several direct intervention trials testing whether lowering cholesterol levels does in fact decrease coronary heart disease risk. These trials were done years ago and each had flaws but, as reviewed in Chapter 3, pp. 48–56, in three of them the reduction in risk was impressive (21–23). Moreover, the magnitude of the effects on cardiovascular events as a function of the magnitude of the drop in blood cholesterol in those early studies matches very closely that relationship in the later drug treatment trials. In short, while there may be minor exceptions, it appears that it is primarily the drop in cholesterol level per se that determines the degree of prevention, no matter which diet or which drug is used to achieve that drop (see Figure 7.3).

A major reason people have questioned the importance of dietary fat as a determinant of blood cholesterol level is that in many epidemiologic studies within a given population (e.g. the Framingham population), no significant correlation has been found. As nicely pointed out by Jacobs et al. (24;25), the wide variability in diet responsiveness from one individual to the next, discussed above, might be expected to dilute out the correlation. Yet every individual in that population would probably show the same *qualitative* response to dietary saturated fat as the subjects in the open Diet–Heart Feasibility Study or in the Ahrens-style metabolic ward study (see Chapter 3, pp. 44–46 and Table 3.1).

"DIETARY CHOLESTEROL INTAKE IS IRRELEVANT TO CORONARY HEART DISEASE RISK"

The importance of dietary cholesterol as a determinant of blood cholesterol levels has probably been overstated; certainly saturated fat intake is quantitatively more important. However, cholesterol intake is relevant. As discussed in Chapter 3, pp. 47–48, if an individual is already on a diet that provides more than 500 mg/day of cholesterol, adding more cholesterol will not increase the blood cholesterol level much further. But if an individual is initially on a cholesterol-free diet, increasing cholesterol intake to 300 mg/day can increase blood cholesterol by as much as 30–40 mg/dl. The typical American diet provides more than 500 mg cholesterol per day so small changes may have little effect. However, on balance, high dietary cholesterol intake is associated with adverse effects on the total cholesterol/HDL ratio (26).

"MOST HEART ATTACKS OCCUR IN PEOPLE WITH NORMAL CHOLESTEROL LEVELS"

Here the key lies in the definition of "normal." At one time any level below 280 mg/dl, the 95th percentile value, was considered "normal." Then it was shown that coronary heart disease risk increases with blood cholesterol level over the full range of values and that reducing levels to 200 or less could reduce coronary heart disease risk significantly. Now of course we know that lowering LDL cholesterol to 70 mg/dl (total cholesterol ca. 150 mg/dl) is more effective than lowering it to just 100 mg/dl (total cholesterol ca. 180 mg/dl). The potential fallacy of equating "normal" with "healthy" or "desirable" is discussed in Chapter 3, pp. 31–33. Of course some heart attacks do occur in people with relatively low blood cholesterol levels because blood cholesterol is not the sole arbiter of coronary heart disease risk. A cigarette smoker is at high risk even if his or her blood cholesterol is on the low side; someone with severe hypertension may have a heart attack despite a low cholesterol value; someone with a low HDL level may actually have a rather low total blood cholesterol and a low LDL cholesterol but he/she is at high risk. So, yes, *some* heart attacks occur in people without a very high blood cholesterol level but certainly not "*most*".

"BLOOD CHOLESTEROL LEVEL DOESN'T MATTER ANY LONGER IN OLD AGE"

This notion arose initially from observations that the correlation between blood cholesterol level and coronary heart disease risk was weaker in the elderly. Physicians tended not to treat the elderly aggressively. However, the large-scale statin studies show that intensive treatment in the elderly (over 65–70) reduces coronary heart disease risk *to almost the same extent as it does in the younger population*. Because the number of events per 1,000 population is much higher in the elderly, the absolute number of events prevented is actually greater and there is thus even more reason to treat.

"NOT ALL SCIENTISTS ACCEPT THE LIPID HYPOTHESIS"

As early as 1978, a poll of 211 scientists working on one or another aspect of atherosclerosis research showed that of 189 responding 185 accepted that cholesterol was connected to coronary heart disease, two were uncertain and two said no (27). Now that the statin results are in (Chapter 8), the hypothesis is established beyond doubt. There remain questions about when to treat and how aggressively to treat but no one with an open mind can any longer question the basic hypothesis. In a recent review (28), Thompson, Packard, and Stone summed it up this way: "The 'lipid hypothesis' is now universally recognized as a law." Possibly that statement is a little too strong, but not much so.

"ATHEROSCLEROSIS IN RABBITS AND OTHER ANIMAL MODELS HAS NO RELEVANCE TO THE HUMAN DISEASE"

The case for the relevance of experimental atherosclerosis is made in Chapter 2. The argument that Anitschkow's cholesterol-fed rabbits were irrelevant because of the unnatural diet they were forced to eat was laid to rest by the discovery of the Watanabe rabbit. Yoshio Watanabe reported in 1980 his discovery of a rabbit that had blood cholesterol levels over 600 mg/dL on a regular chow diet! There was severe atherosclerosis and even skin xanthomas (29). It turned out that these animals had exactly the same mutation in the LDL receptor gene found in some patients with familial hypercholesterolemia (30).

"LOWERING YOUR BLOOD CHOLESTEROL LEVEL WILL NOT INCREASE YOUR LONGEVITY"

Life expectancy refers to the overall average number of years people in a certain population are likely to live. Taylor et al. (31) calculated, using a model based on the extent of cholesterol lowering achievable with diet, that life expectancy would at most be increased by a year in the high-risk group. That doesn't sound very impressive. Consider, however, that if you could get men to stop smoking (which would drastically reduce lung cancer deaths), you would only increase life expectancy (in Canada) by 3–4 years (32). However, that would be saving 4,000,000 person-years of life! Manton et al. (33) suggested that if all cancers were eliminated in the USA, life expectancy would only increase by 2.4 years. However, in the subgroup stated to die from cancer, life expectancy would increase by 13.7 years. Life expectancy is not the way to judge the value of the intervention. Lives *are* saved. Not everyone's, but if yours is one of them, the value is clear. Additionally, preventing a heart attack has a huge impact on quality of life.

REFERENCES

1. 1984. The Lipid Research Clinics Coronary Primary Prevention Trial results. I. Reduction in incidence of coronary heart disease. *JAMA* 251:351–364.

2. 1984. The Lipid Research Clinics Coronary Primary Prevention Trial results. II. The relationship of reduction in incidence of coronary heart disease to cholesterol lowering. *JAMA* 251: 365–374.

3. Frick, M.H., Elo, O., Haapa, K., Heinonen, O.P., Heinsalmi, P., Helo, P., Huttunen, J.K., Kaitaniemi, P., Koskinen, P., Manninen, V., Maenpaa, H., Malkonen, M., Manttari, M., Norola, S., Pasternack, A., Pikkaraineen, J.,Romo, M., Sjoblem, T., and Nikkila, E.A. 1987. Helsinki Heart Study: primary-prevention trial with gemfibrozil in middle-aged men with dyslipidemia. Safety of treatment, changes in risk factors, and incidence of coronary heart disease. *N Engl J Med* 317:1237–1245.

4. Collins, R., Keech, A., Peto, R., Sleight, P., Kjekshus, J., Wilhelmsen, L., MacMahon, S., Shaw, J., Simes, J., Braunwald, E., Buring, J., Hennekens, C., Pfeffer, M., Sacks, F., Probstfield, S., Yusuf, S., Downs, J.R., Gotto, A., Cobbe, S., Ford, I., and Shepherd, J. 1992. Cholesterol and total mortality: need for larger trials. *Br Med J* 304:1689.

5. Wysowski, D.K. and Gross, T.P. 1990. Deaths due to accidents and violence in two recent trials of cholesterol-lowering drugs. *Arch Intern Med* 150:2169–2172.

6. 1971. Trial of clofibrate in the treatment of ischaemic heart disease. Five-year study by a group of physicians of the Newcastle upon Tyne region. *Br Med J* 4:767–775.

7. 1971. Ischaemic heart disease: a secondary prevention trial using clofibrate. Report by a research committee of the Scottish Society of Physicians. *Br Med J* 4:775–784.

8. Dewar, H.A. and Oliver, M.F. 1971. Secondary prevention trials using clofibrate: a joint commentary on the Newcastle and Scottish trials. *Br Med J* 4:784–786.

9. Scandinavian Simvistatin Survival Study Group. 1994. Randomised trial of cholesterol lowering in 4444 patients with coronary heart disease: the Scandinavian Simvistatin Survival Study (4S). *Lancet* 344:1383–1389.

10. Shepherd, J., Cobbe, S.M., Ford, I., Isles, C.G., Lorimer, A.R., Macfarlane, P.W., McKillop, J.H., and Packard, C.J. 1995. Prevention of coronary heart disease with pravastatin in men with hypercholesterolemia. West of Scotland Coronary Prevention Study Group. *N Engl J Med* 333:1301–1307.

11. 2002. MRC/BHF Heart Protection Study of cholesterol lowering with simvastatin in 20,536 high-risk individuals: a randomised placebo-controlled trial. *Lancet* 360:7–22.

12. Jacobs, D., Blackburn, H., Higgins, M., Reed, D., Iso, H., McMillan, G., Neaton, J., Nelson, J., Potter, J., Rifkind, B., Rossouw, J., Shekelle, R., and Yusuf, S. 1992. Report of the Conference on Low Blood Cholesterol: Mortality Associations. *Circulation* 86:1046–1060.

13. Hulley, S.B., Walsh, J.M., and Newman, T.B. 1992. Health policy on blood cholesterol. Time to change directions. *Circulation* 86:1026–1029.

14. Lewis, B., Paoletti, R., Tikkanen, M.J. 1993. *Low Blood Cholesterol: Health Implications*. Current Medical Literature, Ltd, London:1–83.

15. Iribarren, C., Reed, D.M., Chen, R., Yano, K., and Dwyer, J.H. 1995. Low serum cholesterol and mortality. Which is the cause and which is the effect? *Circulation* 92:2396–2403.

16. Steinberg, D., Grundy, S.M., Mok, H.Y., Turner, J.D., Weinstein, D.B., Brown, W.V., and Albers, J.J. 1979. Metabolic studies in an unusual case of asymptomatic familial hypobetalipoproteinemia with hypolphalipoproteinemia and fasting chylomicronemia. *J Clin Invest* 64:292–301.

17. Kinsell, L.W., Partridge, J., Boling, L., Margen, S., and Michael, G. 1952. Dietary modification of serum cholesterol and phospholipid levels. *J Clin Endocrinol Metab* 12:909–913.

18. Ahrens, E.H., Jr., Blankenhorn, D.H., and Tsaltas, T.T. 1954. Effect on human serum lipids of substituting plant for animal fat in diet. *Proc Soc Exp Biol Med* 86:872–878.

19. Beveridge, J.M., Connell, W.J., and Mayer, G.A. 1956. Dietary factors affecting the level of plasma cholesterol in humans: the role of fat. *Can J Med Sci* 34:441–455.

20. 1968. The National Diet–Heart Study Final Report. *Circulation* 37:I1–428.

21. Leren, P. 1966. The effect of plasma cholesterol lowering diet in male survivors of myocardial infarction. A controlled clinical trial. *Acta Med Scand Suppl* 466:1–92.

22. Miettinen, M., Turpeinen, O., Karvonen, M.J., Elosuo, R., and Paavilainen, E. 1972. Effect of cholesterol-lowering diet on mortality from coronary heart-disease and other causes. A twelve-year clinical trial in men and women. *Lancet* 2:835–838.

23. Dayton, S., Pearce, M.L., Goldman, H., Harnish, A., Plotkin, D., Shickman, M., Winfield, M., Zager, A., and Dixon, W. 1968. Controlled trial of a diet high in unsaturated fat for prevention of atherosclerotic complications. *Lancet* 2:1060–1062.

24. Jacobs, D.R., Jr., Anderson, J.F. and Blackburn, H. 1979. Diet and serum cholesterol: do zero correlations negate the relationship? *Am J Epidemiol* 110:77–87.

25. Blackburn, H. and Jacobs, D. 1984. Sources of the diet–heart controversy: confusion over population versus individual correlations. *Circulation* 70:775–780.

26. Weggemans, R.M., Zock, P.L., and Katan, M.B. 2001. Dietary cholesterol from eggs increases the ratio of total cholesterol to high-density lipoprotein cholesterol in humans: a meta-analysis. *Am J Clin Nutr* 73:885–891.

27. Leren, P., Askevold, E.M., Foss, O.P., Froili, A., Grymyr, D., Helgeland, A., Hjermann, I., Holme, I., Lund-Larsen, P.G., and Norum, K.R. 1975. The Oslo study. Cardiovascular disease in middle-aged and young Oslo men. *Acta Med Scand Suppl* 588:1–38.

28. Thompson, G.R., Packard, C.J., and Stone, N.J. 2002. Goals of statin therapy: three viewpoints. *Curr Atheroscler Rep* 4:26–33.

29. Watanabe, Y. 1980. Serial inbreeding of rabbits with hereditary hyperlipemia (WHHL-rabbit). *Atherosclerosis* 36:261–268.

30. Kita, T., Brown, M.S., Watanabe, Y., and Goldstein, J.L. 1981. Deficiency of low density lipoprotein receptors in liver and adrenal gland of the WHHL rabbit, an animal model of familial hypercholesterolemia. *Proc Soc Natl Acad Sci U.S.A.* 78:2268–2272.

31. Taylor, W.C., Pass, T.M., Shepard, D.S., and Komaroff, A.L. 1987. Cholesterol reduction and life expectancy. A model incorporating multiple risk factors. *Ann Intern Med* 106:605–614.

32. Grover, S.A., Gray-Donald, K., Abrahamowicz, M., and Coupal, L. 1994. Life expectancy following dietary modification or smoking cessation. Estimating the benefits of a prudent lifestyle. *Arch Intern Med* 154:1697–1704.

33. Manton, K.G., Patrick, C.H., and Stallard, E. 1980. Population impact of mortality reduction: the effects of elimination of major causes of death on the 'saved' population. *Int J Epidemiol* 9:111–120.

GLOSSARY

Adventitia: the outermost layer of the large arteries.

Angina pectoris: chest pain due to an inadequate flow of blood in one or more of the arteries (coronary arteries) feeding the musculature of the heart (myocardium). It is an indication that the vessel is partially obstructed by an atherosclerotic lesion protruding from the wall. Often brought on by physical exertion that increases the oxygen requirements of the myocardium.

Angiography: visualization of the coronary arteries using a narrow catheter introduced into the femoral artery and threaded up to the heart to inject x-ray contrast medium. This procedure detects partial or complete obstruction of the coronary arteries but does not assess the volume or extent of the lesions *within* the artery wall. Frequently the artery reshapes itself as the atherosclerotic lesion grows so that blood flow is only minimally impaired yet the surface of the lesion is a potential site for thrombosis and consequent myocardial infarction.

Angioplasty: mechanical dilatation of a narrowed segment of a coronary artery using a long, narrow catheter introduced into the femoral artery at thigh level and threaded up to the heart. A balloon attached to the tip of the catheter can be expanded within the narrowed artery to dilate it and restore normal blood flow.

Arteriosclerosis = atherosclerosis (qv).

Atheroma: any of the arterial lesions associated with atherosclerosis.

Atherosclerosis = hardening of the arteries, now used synonymously with "arteriosclerosis": a chronic disease of the arteries that worsens progressively with age and can ultimately lead to heart attacks and/or strokes and/or deterioration of blood flow to the legs (peripheral vascular disease, PVD).

Bad cholesterol: shorthand for the amount of cholesterol contained in LDL (low density lipoprotein). This is an unfortunate colloquialism meant to connote that LDL as a lipoprotein somehow causes atherosclerosis. Of course the cholesterol molecules in LDL are not different in any way from the cholesterol molecules in HDL or IDL or anywhere else. Cholesterol is only one of many different molecular components in LDL but the cholesterol content is what the laboratory measures and it is proportional to the concentration of LDL particles in the blood or plasma. It is the LDL as a lipoprotein that is bad, not the cholesterol in it. Like other lipoproteins, LDL particles are huge – 10,000 times as large as sugar molecules – and made up of thousands of lipid molecules and one large protein (apoprotein B).

Cholesterol: a major lipid component of lipoproteins and of cell membranes throughout the body. There is no dietary requirement for it because endogenous biosynthesis provides all the cholesterol the body needs. Cholesterol is the obligatory precursor for bile acids and for steroid hormones. Cholesterol accumulation in the artery wall is the hallmark of atherosclerosis.

Cholesterol ester: cholesterol that is combined chemically with another lipid, a fatty acid. When cells have taken up extra cholesterol, they store it in the form of cholesterol esters, because in that form it is less toxic than free cholesterol.

Chylomicrons: huge lipoproteins made in the lining of the intestine during the absorption of a fat-containing meal. Fat makes up 90 percent or so of the mass of the chylomicron.

Coronary artery: any of the arteries bringing oxygen and nutrients to the myocardium. There are two major branches, right and left. The latter branches in turn to give rise to the left anterior descending artery.

Coronary bypass surgery: restoration of blood flow to the myocardium by using a segment of a vein, usually taken from the leg, and used to divert blood around (bypass) an obstructed segment of a coronary artery (coronary artery bypass graft, CABG).

Coronary heart disease (CHD): atherosclerosis affecting the major arteries feeding the myocardium. In the early decades of life there are no symptoms or signs of disease because the lesions are not obstructing blood flow and the surface of the vessels do not yet predispose to the formation of blood clots (subclinical CHD). In later decades, blood flow may be reduced sufficiently to cause chest pain (angina pectoris), especially on exertion. Ultimately, a blood clot may form on the surface of an advanced atherosclerotic lesion, precipitating myocardial infarction.

Dyslipidemia: a broad term for any abnormality in the pattern of lipoproteins in the blood. It includes both increases and decreases in concentrations of any of the lipoproteins and/or abnormalities in their chemical composition.

Endothelium: the one-cell thick layer of cells that lines the blood vessels. The normal endothelium inhibits blood clotting (thrombosis); damaged endothelium may not, especially if a part of the endothelium is torn away. This allows blood access to the tissue beneath, which contains tissue factor, a potent initiator of blood clotting.

Fat: in biochemistry and nutrition, used synonymously with triglyceride (qv). Body fat, such as subcutaneous fat, is up to 90 percent stored triglyceride. Fats are lipids by definition – because of their limited solubility in water and ready solubility in alcohol, ether, or other organic solvents. However, not all lipids are fats; the term "lipids" includes cholesterol, phospholipids, and many, many other compounds that have limited solubility in water.

Fatty acids: these chemical compounds make up the bulk of the fats in our diet. Three fatty acid molecules are connected to one glycerol molecule; hence, the name, triglycerides. Fatty acids consist of long strings of carbon atoms attached to each other with an acidic grouping (carboxyl group) at one end of the string.

Fatty streak: the earliest visible lesion of atherosclerosis, consisting of lipid deposits just beneath the arterial endothelium. Much or most of it is initially contained within foam cells, so-called because the excess lipid in them (mainly cholesterol esters) is stored in large droplets that give the cells a foamy appearance under the microscope. Most foam cells are derived from blood monocytes that have penetrated through the endothelial layer and taken up residence in the subendothelial space; some foam cells are derived from arterial smooth muscle cells. At first the fatty streak does not protrude into the lumen, so it does not obstruct blood flow; the endothelium overlying it is intact, so there is no danger of thrombosis. It is therefore a benign lesion, i.e. clinically silent. Fatty streaks begin in childhood and become steadily more prevalent. By age 30 as much as 30 percent of the aortic surface is covered by fatty streaks in normal individuals and the figure is higher for those with hypercholesterolemia. The most important thing to recognize is that the later lesions, which are clinically relevant, derive from these fatty streak lesions.

Fibrous plaque: an atherosclerotic lesion that has progressed beyond the fatty streak stage and now shows, in addition to foam cells, growth of smooth muscle cells, deposition of fibrous tissue, and lipid deposits between the cells. It can progressively enlarge and partially impede blood flow (stenosis). It often enlarges within the wall of the vessel so that the obstruction to blood flow may be minimal even though the lesion may be rather large and its irregular surface represents a potential site for thrombosis (qv).

Genotype: the normal or mutant genes of a given individual that control the proteins or functions of interest.

Good cholesterol: the amount of cholesterol contained in HDL (high density lipoprotein). An unfortunate colloquialism to recognize that HDL somehow confers protection against atherosclerosis. Of course the cholesterol molecules in HDL are not different in any way from the cholesterol molecules in LDL or IDL or anywhere else.

HDL: high density lipoprotein. Hydrated density >1.063. HDL picks up excess cholesterol from tissues (including the arteries) and carries it back to the liver, which can excrete it in the bile either as such or after converting it to bile acids. High levels of HDL are associated with lower risk of atherosclerosis and its complications.

Heart attack: *see* myocardial infarction.

Heterozygous/homozygous: genes occur in pairs (alleles). If one allele is abnormal and the other is normal, the individual is heterozygous for that gene. If both alleles are mutated, and mutated in the same way, that individual is homozygous. For example, an individual with familial hypercholesterolemia might have mutations in both alleles for the LDL receptor (homozygous) and have extraordinarily high blood cholesterol levels (over 600 mg/dl); or they might have one normal allele, making normal

LDL receptors, and one mutant allele (making little or no receptor). The latter patients are heterozygous. They have blood cholesterol levels intermediate between the normal and the levels in homozygous patients (about 350 mg/dl).

HMGCoA reductase: the enzyme in the pathway for cholesterol biosynthesis that determines the rate at which cholesterol will be made, i.e. the rate-limiting enzyme. This is the enzyme inhibited by the statins.

Hypercholesterolemia: elevation of blood cholesterol level above "normal." As discussed in detail in Chapter 3, the definition of "normal" is by no means straightforward. At one time, any value for blood cholesterol below 300 mg/dl was considered "normal." Eventually it became clear that even people with values near 200 mg/dl were more likely to have heart attacks than those with lower values, and that treatment to lower the cholesterol level below 200 reduced the incidence of heart attacks.

Hyperlipoproteinemia: relevation of the blood concentration of any one or more of the lipoprotein fractions.

IDL: intermediate density lipoproteins. Hydrated density 1.006–1.019. These are derived from VLDL by the action of an enzyme (lipoprotein lipase) that breaks down the triglycerides of VLDL, making the VLDL progressively smaller. IDL represent the "remnants," i.e. the smaller lipoprotein particles remaining after most of the triglycerides have been broken down and taken up by tissues. These remnant lipoproteins are believed to be highly atherogenic.

Intima: the innermost layer of the large arteries.

LDL: low density lipoprotein, the major carrier of cholesterol in the blood.

Hydrated density 1.019–1.063. High LDL levels are associated with increased risk of atherosclerosis and its complications.

Lipids: compounds that have extremely limited soluble in aqueous media but dissolve readily in alcohol, ether, or other organic solvents.

Media: the bulk of the arterial wall, made up mainly of smooth muscle cells and matrix.

Meta-analysis: in statistics, a formal way to pool the information derived from multiple trials of the same hypothesis in reaching a decision regarding its validity.

Monoclonal: deriving from a single parent cell of origin. The monoclonal hypothesis for atherogenesis postulated that all the smooth muscle cells in an atherosclerotic lesion arose from multiplication of a single cell of origin.

Monocyte/macrophages: monocytes, white blood cells containing a single nucleus, exit from the blood stream and penetrate into the artery wall through spaces between endothelial cells. Once out of the blood stream they differentiate, i.e. change their patterns of gene expression. They are then designated macrophages or monocyte/macrophages. Most of the lipid in fatty streaks is found in macrophages.

Monogenic: refers to a disease caused by mutation of just one specific gene. All disease manifestations can be traced to the abnormality in the functioning of that one gene, as in the case of familial hypercholesterolemia.

Monounsaturated fat: a food fat relatively rich in monounsaturated fatty acids (e.g. olive oil).

Monounsaturated fatty acid: a fatty acid in which one, and only one, pair of carbon atoms has each lost a hydrogen atom. Those two carbon atoms now have two chemical bonds linking them instead of just one – hence, a "double bond."

Myocardial infarction = heart attack: death of a portion of the cardiac muscle (myocardium) due to obstruction of one of the major branches of the arteries feeding the myocardium. Most often due to a sudden obstruction caused by a blood clot forming at the surface of an artery affected by advanced atherosclerosis. In a high percentage of cases this may occur without any previous warnings of disease. In other cases, it occurs in a patient who has had angina pectoris because of partial obstruction of a coronary artery. In 25–35 percent of cases, the first attack is fatal.

Myocardium: the muscular part of the heart, responsible for the pumping of blood to the lungs and the rest of the body.

NIH: National Institutes of Health, the major federal establishment for support and encouragement of biomedical research in the USA. It is the major source of funding for biomedical research at universities all over the U.S.A.

p-value: a statistical parameter that assesses the probability that an observed difference in an experiment might be due to chance alone. For example, let us say that a new drug is tested in 1,000 men with coronary heart disease, using another 1,000 men similarly affected by coronary heart disease but not given the drug as a control group (the placebo group). Over a 5-year period, 80 of the men getting the drug die of a heart attack; 100 of the men not getting the

drug die of heart attack. That difference (a 20 percent decrease in risk) might have occurred by chance if the men assigned to the two groups originally were not at the same relative risk (even though assignment to the groups was done in a random fashion). How likely is it that chance alone accounts for the observed difference? If the p-value is 0.01, that implies that there is only a 1 percent chance that the observed difference is due to such a misdistribution of the participants. In other words, it is highly likely (99 percent probable) that the drug is actually effective. Arbitrarily, results are considered "statistically significant" if p is less than 0.05.

PDGF = platelet-derived growth factor (qv).

Phenotype: the observable traits or properties resulting from the expression of a gene or genes. For example, the phenotype of a patient with homozygous familial hypercholesterolemia could include cholesterol deposits in the skin and a blood cholesterol level over 500 mg/dl.

Plasma: the clear, cell-free liquid portion of blood that has been treated with an anticoagulant and prevented from clotting. It differs from serum in that it still contains all the factors needed for clotting.

Platelet-derived growth factor (PDGF): a small protein found originally in blood platelets but later found to be present in many different cell types. It is a growth factor for many cells in culture.

Polyunsaturated fat: a fat relatively rich in polyunsaturated fatty acids (e.g. corn oil, safflower oil).

Polyunsaturated fatty acid: a fatty acid in which two or more pairs of the carbon atoms are joined to each other by double bonds (see monounsaturated fatty acid).

Primary prevention: treatment intended to prevent a first heart attack.

Saturated fat: a food fat relatively rich in saturated fatty acids (e.g. all animal fats, cream, butter, lard, coconut oil).

Saturated fatty acid: a fatty acid in which all of the carbon atoms along the chain (excepting the carbon at the end, which is attached to oxygen) are attached to each other or to hydrogen atoms; there are no chemically reactive sites left open.

Secondary prevention: treatment intended to prevent further heart attacks in a patient who has already suffered one or more.

Serum or plasma lipoproteins: huge molecules made up of protein and lipids (cholesterol, triglycerides, phospholipids). Essentially all of the lipids in the blood are transported in these complexes. The protein component is what allows the lipids, which are otherwise barely soluble in an aqueous environment, to stay in solution in the blood.

Serum: the clear, cell-free liquid portion of blood that has been allowed to clot. The cellular components are carried down with the clot along with the clotting factors.

Smooth muscle cell: the major cell type in the walls of the large arteries. These cells can contract and relax, causing the channel for blood flow to decrease or to increase (vasoconstriction or vasodilatation, respectively).

Statistically significant: *see p*-value.

Statins: a class of drugs that lower blood cholesterol by inhibiting its synthesis at the key step involving the enzyme HMGCoA reductase. These are the most effective drugs now available for lowering blood cholesterol and they are remarkably safe.

Stenosis: narrowing of the blood flow channel in an artery, as in an atherosclerotic coronary artery.

Thrombosis: the formation of a clot in a blood vessel. If the clot is large enough to obstruct blood flow, the tissue fed by the thrombosed artery dies. If the vessel is one of the coronary arteries, the result is myocardial infarction (qv).

Triglyceride: a compound made up of three fatty acid molecules attached to glycerol.

VLDL: very low density lipoprotein, the major carrier of triglycerides in the blood. Hydrated density <1.006. The primary lipoprotein synthesized and secreted by the liver. Occur over a wide range of sizes in the plasma (*see* IDL).

Xanthomatosis: deposits of excess cholesterol in the skin (elbows, knees, buttocks) or in the tendons. Seen in patients with extremely high blood cholesterol levels.

INDEX

Printed and bound by CPI Group (UK) Ltd, Croydon, CR0 4YY

03/10/2024

01040415-0003